Rivet Boy

Barbara Henderson

First published in 2023 by Pokey Hat

Pokey Hat is an imprint of Cranachan Publishing Limited

Copyright © Barbara Henderson 2023

The moral right of Barbara Henderson to be identified as the author of this work has been asserted by her in accordance with the Copyright, Designs and Patents Act, 1988.

All rights reserved.

No part of this publication may be reproduced, stored in a retrieval system or transmitted in any form or by any means, electronic, mechanical, photocopying, recording or otherwise, without prior permission of the publisher.

ISBN: 978-1-911279-22-8

eISBN: 978-1-911279-25-9

Cover Photo Forth Bridge © Shutterstock / Richie Chan

Cover Photo Squirrel © Shutterstock / Taekhmister

Interior Illustrations © Sandra McGowan

Additional photos are out of copyright unless otherwise stated

cranachanpublishing.co.uk

@cranachanbooks

cranachan

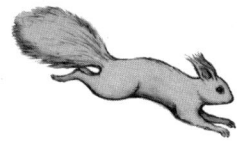

To Frank Hay,
without whose
infectious enthusiasm
this book could not
have been written

Prologue

FIFE MAN KILLED IN SHOCKING ACCIDENT IN AUSTRALIA

A Fife emigrant lost his life in Australia on Wednesday 17th May 1876.

John Nicol, 32, had only recently arrived in Melbourne to start a new life in the colony, and to prepare for the arrival of his wife Janet and their unborn child from Scotland. However, tragedy struck when the unfortunate labourer was crushed by a falling ship boiler just days after his arrival. A loose rivet was found to be the cause of the accident.

Following an outpouring of sympathy for the pregnant widow across the ocean, charitable Australians set up a fund to support the family until Mrs Nicol remarries or until the child, if a son, is old enough to be a breadwinner.

1

Beggars Cannae Be Choosers

Twelve years later, September 1888

I want to tear it, stamp on it, slash it. All my rage and regret, directed at a shirt. And it's not even mine.

Mother and the ironing board have disappeared into a cloud of urgent steam. An empty cardboard box of Colman's Starch lies discarded on the floorboards until she kicks it aside and re-emerges rather dramatically, holding the shirt aloft, gleaming white. It's cold in our flat, so the moisture in the air settles on the ironing board and every other surface almost immediately, including her forehead. She looks almost feverish.

'Put it on!' urges my grandfather, from his chair in the corner.

'Yes, Dey,' I answer, barely audibly. Never has a borrowed shirt been buttoned more hesitantly, I'm sure. I feel heartily sorry for myself, but of course I hold my tongue.

Dey struggles to his feet. 'Now, less of that whingin face, John. I'm tellin ye, beggars cannae be choosers. It's no use now, all that fancy stuff they teach ye at school. That's fur bairns. You're old enough to help yer mother now.'

My Dey turns his face away. His lower lip trembles above his grey-white stubble.

Tears are pricking my eyes. I wish they weren't, but they are, and my grandfather has seen them, which is worse. I take a deep, deep breath to keep them at bay. *I wish they would understand. I have friends at school. The school, with its inkwells and charts and maps of the Empire curling off the wall. That's where I belong. I am not ready to be a working man.*

Mother pulls on her coat, clasps her scuffed bag under her arm and gives me the warning look. 'John, we need to leave this minute, or else we'll miss the train.'

I close the mouth I had opened to protest one last time. She brushes a strand of loose hair from her forehead and takes wee Helen's hand. My little sister has only just learned to walk. She giggles when I pull a face, and Dey smiles, which makes me feel a bit better.

Mother pushes me towards the door. 'Mind, don't talk once we get there, John. Leave that to me. And for goodness' sake…' She stops and attempts to flatten the hair on the back of my head which shoots up like bolts of straw-coloured lightning.

'I'll wear the cap, Mother.' I can't help my freckles, but I can hide my unruly hair! Heading towards the door, I reach for the hook beside the mantlepiece where a faded picture of the man I have never met has been my lifelong companion. She has seen me glance at it.

'Your Dey is right,' she says under her breath as she bundles me out of the door. 'You're old enough to be a worker like your father was, John, and your family needs you now.' She pulls the cap firmly over my ears. 'Don't make it harder for your Dey than it needs to be. He hates to see you so upset. Unfortunately, he is right about this too: beggars can't be choosers. Whether we like it or not, there is no other way.'

Helen toddles into our neighbour Bessie's house and Mother and I pick up our pace to the station. I don't feel myself, with the starch-smell rising from the stiff shirt and wearing our neighbour's shiny shoes. We can hear the whistle of the train from afar.

Everyone, and I mean *EVERYONE,* on the train is a brigger. The air in the carriage is thick with curses and damp with sweat. We jiggle, tightly packed in the rattling compartment, while the steam curls past the window, revealing the green fields, the patchwork of trees and the glittering water of the Firth of Forth in the distance. I know some of the faces from the market, or from church—and some of the younger ones used to attend my school, thin fluff growing on their boyish chins. That one over there is David Clark, only a year older than me. Our eyes meet for a moment.

The nearer we get to the shore, the tighter the buttons at my throat feel. Beside the muscled and scarred men around me, I feel no more grown up than my little sister.

In the jostle to disembark, I can barely stay upright. We are swept towards the giant metal structure of the bridge rising from the water, so much bigger than I expected. Angular limbs of steel stretch their way towards the sky in intricate lattice patterns: three giant diamond shapes separated by water and, once you look closely, crawling with movement. Seeing the tiny figures near the top lands like a punch in my stomach. Those dots are briggers— and despite what I may wish, I am destined to join them. Like a splinter in the skin, the bridge somehow doesn't seem to belong in the smooth surface of the water. Even the birds circle its towers in outrage.

Stone viaducts on both sides of the river lead towards the construction site. It is hard to imagine trains steaming along them and over the Forth estuary from Edinburgh to Fife, and yet, that is what they tell us the bridge is for. The rhythmic sound of the waves is punctuated by a million hammers and hollers while ships of all kinds zigzag the sunlit surface of the Forth below: ferries, fishing boats, pleasure vessels, lifeboats and brigger transport.

'Come, John.' Mother receives the usual winks and

whistles. Despite being in her thirties, she is what Dey would call 'a looker', with her easy grace and her pleasant demeanour, and a voice like velvet. But these men don't know about her rock-hard resolve. No one messes with Janet Nicol, make no mistake. Least of all me.

She stops to ask a curly-haired brigger for directions to Mr Arrol's office, but the youth smiles at her broadly and booms: 'Mr Arrol's office is on the other side, Miss, and he's not always on site in any case. But you're in luck today.'

'Mrs,' my mother corrects.

'Mr Arrol happens to be right here on this side of the Forth this morning. See that distinguished gentleman with the blue neckerchief? That's him down there on the pier, talking to the two bridge designers. Mind, I wouldn't interrupt them just now, Miss, they'll be talking about something important. What's it concerning?'

'Please call me Mrs,' she corrects once more. 'My son here is in need of employment, so I resolved to come and speak to Mr Arrol.'

The young man laughs. 'You want to bother Mr Arrol with that? Listen Miss, Mr Arrol is in charge of this whole bridge construction. Do you really think he will care about giving a boy work? Let me tell you, Mr Arrol is far too important to bother with the hiring; he's got people for that sort of thing. You can write to the offices in Queensferry, perhaps?'

Mother's face falls.

Before I can control them, my legs carry me right down the pier towards the three gentlemen in earnest conversation, leaning together over a very detailed drawing which distracts me for a moment. It is so very beautiful, though I do not understand the patterns of lines and numbers. Above me, seagulls continue to scream. I choose to take it as encouragement.

'John!' Mother hisses behind me, trying to catch up in her hand-me-down heels, but my mind is made up.

2

Luck Favours the Brave

'Mr Arrol, sir? May I speak with you?'

All three men raise their heads and look at me and I'm sure I can hear an involuntary groan from Mother behind me.

He glances at his colleagues and takes a step towards me, looking me up and down. 'Indeed, I am William Arrol. How can I help you?' He is broad and tanned with lively eyes, and wears his authority like a greatcoat, loosely.

Mother interrupts: 'I am so sorry. We'll make an appointment and come back when it is suitable, sir. Come now, John.'

I don't move. 'Please! I'll only be a minute, sir.' My voice is clear now and loud enough to be heard by some of the workers nearby. 'It is a matter of the utmost urgency.'

A small smile creeps onto the man's face. I raise my chin. This man and I are not equals, but I must try to *believe* that we are. If I believe it, he may believe it. *Luck favours the brave*, Dey would say.

'You see, Mr Arrol, we live in a small flat in Dunfermline and I'll be twelve years old tomorrow. My father was killed in an accident overseas before I was even born, and my mother has made ends meet with a widow's allowance and my Dey's work. But now my Dey's old and has got a bad back and knee and can't work anymore and the widow's fund from Australia has run out and I need to be a breadwinner now, so that we won't starve. My wee sister is only two, and my mother is an honest, hardworking soul raising us on her own. Please, sir! It's in your power to help us and give me employment. I may not look it, but I am strong, and second in my class at school—*was* second, that is, until last week when I had to give it up, but I am quick to learn, I really am.'

At that point, my breath runs out. My mother's hand claws into my shoulder to steer me away, but I resist the pull and her urgent whispers, broadening my stance and folding my arms, and looking the gentleman right in the eye.

He still smiles. 'That was quite a speech! You'll make a politician one day! But you know as well as I do that every boy must attend school until he is thirteen. It is the law.'

I expected this, and more than anything, I wish it was true for me.

But Mother has an answer. 'With respect, Mr Arrol, not if he is a breadwinner. He can work on the bridge at twelve if he is a breadwinner.' Mother's soft tone barely carries over building noise.

Mr Arrol hesitates as his two colleagues whisper among themselves. 'You have given this a great deal of thought, I see,' he finally says.

Mother and I stand shoulder to shoulder. Most tools nearby have fallen silent as workers watch our stand-off. I expect Mr Arrol to tell us to begone. He has every right to do it. But there is compassion in the man's eyes, despite the top hat and the heavy coat that mark him out as a man of consequence. Still, he'll boil his eggs in water like the rest of us; we're not so very different. Dey says *we're a' Jock Tamson's bairns*. I hold his gaze.

He points. 'See the rivets, boy? The metal studs holding the structure together? I need boys to work on those. How many rivets do you think a bridge like this will need? If you were second in your class, you might know.'

I didn't expect that kind of question and am flummoxed for a moment. I try to think of the highest number I have ever been taught. 'A hundred thousand, sir?'

His men laugh openly now, but Mr Arrol's face merely crinkles in passing amusement. 'More than six million rivets, boy. Six million. I daresay there is work here for you to do. But it is dangerous work.' A shadow crosses his face.

'I know it is dangerous work, sir. A loose rivet killed my father. Believe me, I am the careful sort. You won't be disappointed in me, I promise, sir.'

'All right, John, that's enough. Leave the gentleman be,' my mother protests weakly behind me, but I am determined to reel this fish in, even though I hate the taste of it. 'When can I start?'

Oh no! Have I gone too far? Rushing out my words, I add: 'Begging your pardon, sir. Please. For the sake of my sister and my ailing Dey, and my poor mother here...'

'Very well! I think we can do something for you,' Arrol says, raising his eyebrows and looking over his shoulder. 'Report to the Queensferry cantilever office Monday morn at seven. You won't require a fancy shirt for that, mind. Comfortable clothing, sturdy boots and a warm jacket are all you need to bring. And in time, a head for heights, I suppose.' He dismisses Mother and me with a nod before turning his attention back to the drawing and the two designers.

We walk back up the pier, adding the clickety-clack of her heels to the noise of the tools. Mother's words wash over me, berating and praising my boldness in equal measure. Passing through the shadow of the giant steel structure on our way back to the train, I can't help glancing up, and up and up and up.

Mr Arrol's words echo in my mind. Now I have a new problem.

The collar of my starched shirt feels tighter and tighter, however much I loosen the buttons. It's true, I now have employment. Dey's old boots and his warm jacket will do.

But what I most definitely lack—is a head for heights.

3

Don't Look a Gift Horse in the Mouth

Now that it's all decided, I feel a strange sense of calm. Perhaps it's the rattling of the train, sifting my worries from my thoughts. For the first time today, I remember that there is tomorrow to look forward to.

In the distance, the rigid skeleton of the bridge is still visible over the hilltop, dwarfing all else in sight. Our train chunters along the Firth and through the wide fields of Fife. As the sun beats down and the wind bends the trees and hedges outside, I detect the flicker of memory in Mother's eyes. I watch her for a while before asking. 'I thought you'd be relieved, Mother. Aren't you?'

She smiles defiantly and keeps her voice light. 'Of course, I am.'

She pauses before continuing. 'I'm just thinking of the opportunities which lie in store for you, John. They'll train you in the workshops in Queensferry where all the materials are made up on the hillside, not up on the bridge itself. The steel sheets are shaped there, cut and drilled, you know, and labelled with chalk. There are tool sheds and iron forges, all manner of progress, it'll be an experience for you, John. And then, come spring…'

Mother doesn't finish her sentence.

I prefer it that way.

I did it. I secured employment and we will be able to eat and pay the rent. That's all that matters.'

I glance across to my mother's face. There it is again, the flicker. It's a passing shadow over her eyes, a brief thought of what could have been if my father had lived, or if our rich neighbour had made good his promise and married her instead of running a mile once my sister was on the way.

I take a deep breath. *Things are not so very terrible, are they? I could talk the hind legs off a donkey, Dey says, and I'm quick thinking. Maybe I'll be able to talk my way out of working on the bridge itself.* I fumble my hand into the pocket of my jacket where a Penny Dreadful is folded up. A sensational story is just what I need to take my mind off it all. I will not allow myself to be sad today; I simply will not have it. *Today is a good day! Next week, I'll be a working man, which makes this a good day! I must hold on to that.*

But when we disembark from the train and walk past the turn-off for the school, I swallow hard. What are the boys in my class doing right now? What game did they play before the bell? Have they even noticed that I am no longer there? Despite my mother beside me, loneliness lunges at my throat. No more reading and arithmetic and thinking about all sorts of interesting ideas. No more chasing my friends in the park after lessons. From now on, my life will be wind and rain, coal, sparks, flames, glowing metal and hammer blows.

When you put it like that, it sounds almost adventurous, and I'll settle for that.

Dey has come all the way down the stair and is waiting for us in the street, with wee Helen clinging on to his good knee. 'Saw you from the windae,' he explains with a shrug, but he stares at Mother until she smiles a tight smile. 'He starts on Monday.'

Dey nods, twice. 'That's good,' he wheezes. It takes him a long time to hobble up the stairs again and I feel sorry for him. I can't imagine what it must be like to feel pain all the time, and I don't particularly want to find out.

'Wait there, Janet!', our neighbour Bessie calls out just as we make to follow Dey up to the flat. 'Hold on, there! The postman came! Look here, for you—all the way from America.'

She holds out a parcel. A parcel! We rarely see the postman in these parts because none of us are high and mighty enough to merit mail. But she is right, the parcel bears an American stamp. Mother approaches it with awe. 'It's from your Aunt Jemima!'

My mother was heartbroken when her favourite

cousin emigrated last year. We had a couple of letters since, but never this, never a parcel.

'Open it, Mother!'

My voice has broken the spell. 'Thank you, Bessie,' Mother whispers and clutches the parcel to her coat as she half runs up the stairs after my Dey. It falls to me to heat the water in the kettle on the range while she reverently removes layer after layer of paper.

'A shawl—look how beautifully she has embroidered it!' Mother models the pretty fabric and looks ten years younger, just like that. There is a rolling sort of toy for Helen, like a duck-shape, and a letter. 'Wait, what's this at the bottom? Oh, John! There is something for you.' Mother half-reads the letter and half-unpacks the parcel with her other hand. 'What's this now? A bible?'

My heart beats a little faster. *A book, for me?* I love the stories I hear in Sunday school, and also the Penny Dreadfuls which my school friends and I used to share by pooling our pennies. However, when she hands me a slim, leather-bound volume and I begin to flick through, something feels wrong.

'Mother, the pages are all blank!'

Dey shuffles over to look, and snorts. 'What's the use o that?'

Mother runs her finger down to the relevant part of the letter and reads out, a little hesitantly because her reading is not good. 'For young John, I have something truly special. It is all the rage here in America: an autograph book. The thing to do is, he should fill it with signatures of famous people, people who he admires in some way. Some autograph collectors build up quite a collection, and some sell theirs and make a great profit. I thought it would be fun for him, and perhaps it will arrive in time for his 12th birthday.'

Mother and Dey stare at each other befuddled, but I leaf through the book again. What celebrities and eminencies am I ever likely to meet? Whose signatures am I going to paste into this book? What use can this gift possibly be, to a boy like me?

Dey looks more confused than ever. 'Americans, eh? What good is that for a craze? Collecting names? I dinnae ken, that's for sure!'

'You always say not to look a gift horse in the mouth, Dey,' I say severely, suppressing a smirk.

'Humph.' With that he sinks onto his favourite chair for a rest while I reluctantly tear myself away from the book. Placing it under my pillow with care, I return to the range just before the kettle whistles.

4

Luck is Loaned, Not Owned

13th September 1888

I wake in a pool of sweat. Seagulls screech, clouds roll and lightning flashes in my dream-darkness. I bite my lip so hard that I taste blood and open my eyes.

A wedge of moonlight shines onto the floor through the gap in the curtain, and the sky is already lightening a little. Dey snores in his little bedroom off the hall, the only bedroom in our flat. Mother shuffles on her mattress in the kitchen alcove and I pull my blanket tight around my shoulders where I lie on the floor. What little light and warmth the range offered has disappeared, and now our flat is shrouded in a ghostly gloom. The second I close my eyes again, I am falling, falling, falling…

'John!' Mother's whisper is directed right into my ear, and I wince away with the fright. 'What in heaven's name is wrong with you? Can we get a wink of sleep around here if you please? You of all people will need it, with starting as a brigger on Monday.'

I can't describe what I am seeing in my mind's eye, so I don't try.

'Sorry, Mother.'

'If you cry out like that again, the wee one will wake, I'm sure of it, and then none of us are going to get any sleep at all—wheesht now!' she whispers.

Her wagging finger is right in front of my nose. I answer by rolling my eyes. She clips me round the head—I didn't think it was that obvious! Gosh, I need to keep my temper in better check.

Turning towards the range and facing the kitchen wall, the sounds of the night envelop me again. Footsteps from the tavern, the hoot of an owl, the faraway rush of the Firth. The rustling.

My eyes ping open. I fear the rats almost as much as I fear heights, but there is no need to worry—it's only a mouse, still a little fluffy. It has emerged from the crack beside the cooker, looking bewildered and disorientated just like me. Scrambling and snuffling, it makes its way along the edge of the wall. Can it tell that I am watching? Lying on the floor, we are the same: too young and too fluffy for the big world around us. On impulse, I sit up and the mouse shoots back into the crack from whence it came. Mother is breathing deeply again. Moving slowly, I raise myself up and tiptoe to the larder. The door doesn't creak much if you lift it a little, and I reach in for a crust of bread. Just a little will do the job. I crouch in front of the crack and place the crumbs on the floor in the shadow of the range. This way it looks like we missed this corner in the sweeping.

Crawling back under the cover, I stretch out and stare at the ceiling, but sleep will not come. Thoughts do though, and worries. Today is my birthday. No one should be sad on their birthday. Besides, I am now the proud owner of a book with empty pages.

I doze until Mother begins to stir in the kitchen.

Thursday, the 13th of September 1888. I am twelve.

I can't help feeling excited. After all, I am going to the new Carnegie Library as a birthday treat. With Dey's bad leg and Mother so busy with the wee one, I am usually needed for chores and errands, but today is different. By late morning, Mother says the words I have been waiting for.

'Right! Now don't say Mrs Nicol does not keep a birthday promise to her only son. You're a man now, John, and you'll be a working man all too soon. I wish it was different, but that can't be helped. Today is your day. You may go to the library; I know you want to. Take out your membership as a birthday treat. Now, I've been in to see the librarian and he assures me there will be no problem. Remember to look after anything you borrow though—we've no money to pay for replacements, mind.'

'I will, Mother.'

I walk up along Priory Lane and St Margaret Street

before turning the corner to reach the Abbot Street entrance to the library. I have often walked past and wondered what a proper library may look like inside. The well-to-do gentlemen who step so confidently up the stone stairs to its imposing door belong to another world. How could I join them? I feel awe, but also peace as I narrow my eyes against the sun, high in the noon sky, and read the words. *Carnegie Public Library*. The letters are chiselled into the pointed arch above the door, rounded, swirling and simply perfect. The carved stone sun beneath the arch seems to be turning a blind eye to me, the boy sneaking into a man's world. Taking the three steps in one giant leap, I am through the door and into the muted light inside. For a moment, it takes my breath away.

How silent it is in here, like a church. I am reminded of the deafening noise of the bridge site. Here one can breathe and think. Dust motes dance in the sunlight from each window. The tiles ooze a comfortable cool, and in the distance, posh shoes tap up and down the polished staircase with its wrought-iron banister. I tiptoe into the lending room and suddenly, stories soar all around me.

Each book spine holds such promise, such secrets. I was young when it opened as a library in 1883, but I can just about remember it as the Commercial Bank as it was before. What a building! A man in a top hat walks past me towards the gentlemen's reading room, clutching a large volume. A young woman in high-heeled boots comes out of the Ladies' Reading Room and glides past me towards the door. Breathing in the book scent greedily, I make my way towards the shelves which creak with the weight of wisdom. Over in the corner, the librarian—a monocled and bearded gentleman—sends his lanky assistant towards the Smoking Room. From his desk, he speaks to his clients in hushed tones. Ordinary men in scuffed shoes walk past corseted ladies with crinolines rustling beneath their skirts, and men with top hats and walking sticks, their watch chains clinking from their waistcoats.

And me.

I must have been in here for almost a quarter-hour before I summon up the courage to reach out and touch a leather-bound copy of a book, just for a second. I pull my hand back and inspect the book for damage, as if my hand could have soiled it. *The world's words are here.* It is amazing to me that anyone may borrow these books and read them for free.

'Are you John Nicol?' a quiet voice says behind me, and I jump.

'I didn't…' I begin my defence, but the librarian's eyes are kind. 'Your mother came in to take out your membership. She said you'd be here alone this afternoon. Not many boys come on their own. Here!' He pulls a slim volume off the shelf and wipes the reddish-brown cover with his sleeve. The golden letters on the spine of the book sing out its title.

'*Treasure Island*. Have you heard of it? It's an adventure story by an Edinburgh writer called Mr Robert Louis Stevenson. I am certain you will love it.' With this he places it into my hands and walks towards the desk. 'Are you coming now?' he whispers and motions over his shoulder and I hurriedly answer the summon.

At the desk he stamps the book with a date below all the others. 'It's popular, this one. It has only been out for a few years,' he smiles and then hands the book to me. 'Do you own any books, John?' he asks.

I laugh, before realising that my response may appear rude. 'No, sir!' I correct myself. 'We don't have things we don't need in our house. Oh wait, I was given a sort of book for my birthday.'

He frowns, but in a perplexed way.

'A sort of book? How can it be a sort of book?'

I look away. 'My aunt sent it from America. It has no words in it at all. She says it's for autogr… auto…'

'Autographs? I have heard of these things. Interesting. May I see this book?'

'It's at home, sir.'

He sinks into deep thought for a moment or two. Then he smiles, as if he had resolved on something secret and wonderful. 'Bring it in the next time you come to the library—when you return *Treasure Island*.'

I stare at him, uncomprehending. 'If you like, sir. It's not much to look at.'

'Just wait and see. If I am not at my desk, ask for Mr Peebles. That's me.' He points at *Treasure Island*, cool and smooth in my hands. 'Now, enjoy this. I look forward to hearing how you liked it. And if you can, come early in the day. It gets terribly busy here at night when the work people are set at liberty. Did you know that on our first day here we issued over 2000 books?' He winks and then laughs at my wide-open mouth. 'Now hurry on, John.'

I close my mouth and mumble an apology to the gentleman who has been waiting behind me. 'Thank you, sir!' I breathe to Mr Peebles. Stepping back from the desk, I open *Treasure Island* with trembling hands and gasp. There is a map, for a start. I love maps! A map is a way of knowing a place without being there, my dominie at school said.

Stroking the spine of the book gently and hardly believing my good fortune, I walk towards the door half-in-a-dream, which is why I nearly collide with the man running into the library.

This feels wrong. The man's heavy bootsteps, his oily hands and his loud voice, in a place so sacred with stories. He makes no effort to keep his voice down.

'There's been another accident on the brig. A boy is dead. A Dunfermline lad, only thirteen years of age. He fell. Dead. Name is David Clark.'

A chill travels up my spine and trickles down my arms. There is rustling as ladies in corsets and men in greatcoats hurry to the door. After a moment rooted to the spot, I follow apace, running. As we spill out onto the corner of Abbot Street and St Margaret's Street, people of all ages and status are gathering around the messenger, clearly a brigger himself. He sounds distressed. 'I'm telling you: I saw him fall with my own eyes. A rivet boy on the brig, nice lad. Lived with his granny in Maygate they say. God rest his soul.'

Some people ask questions, but to my ears everything is a blur until the man speaks again. 'That's right, aye, that's him. He missed his footing around midday, I hear, and fell over 150 feet.'

Every last man and woman winces at that. Bile rises in my throat.

'At least it was quick. He probably didn't feel a thing. Well, there for the grace of God go you and I.' The man walks away, with a handful of gossips pursuing him for more. *David Clark. The boy from school, and the boy on the train. I think of the moment only yesterday when our eyes met in the crowded carriage. Dead. On my birthday.*

And now I am doomed to be a brigger, just like him.
The 13th.
An unlucky day indeed.

5

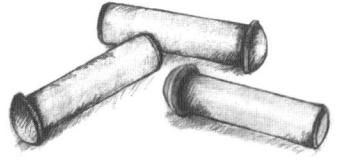

Nothing Ventured, Nothing Gained

No one pays me the slightest heed on the train. Trying not to cough from the pipe smoke in the air, I have pulled the cap low into my face, affording me the chance to steal glances at those around me. Just like last time, every single person on the train is a brigger. Self-consciously, I tug at the overlong sleeves of my Dey's old wool jacket. From the station, I am swept along with the tide of workmen. 'How do I get to the Queensferry side?' I ask a friendly-looking, portly man. He points to the pier. 'See those boats? They shuttle the briggers back and forth all day. Go, lad, or you'll miss this next one!'

Half an hour later, I approach the supervisor's office on the Queensferry side and clear my throat politely until the important man at the desk looks up. He wears a brown tweed jacket, much more expensive than I could ever afford, but his shoes are muddy—he obviously walks among the workmen too.

'Are you looking for your father?' His eyes have

turned back down to his papers. Each word is spoken with calm precision.

'No, sir. I'm here to work.'

That gets his attention. 'I beg your pardon, young man! However, you are too young to work here. If we took every boy who showed up at my office…'

'I'm a breadwinner, sir. My name is John Nicol. I have just turned twelve.'

The man pulls over a ledger from the far end of his desk. I can't help noticing that his inkwell is topped up and shiny, and his steel nib pens are lined up like soldiers beside the neatly stacked papers. Nothing is out of place, from the parting of his silver hair to the moustache framing his thin lips like a theatre curtain. *I'm going to call him MacOrderly in my head, just for fun.*

His buttons are so polished that I can see my reflection in them. Suddenly, I feel ridiculous with my blond-brown straggly hair. Dey's cap is too big for me and keeps sliding over my eyebrows and onto my nose. I must ask Mother to fix it for me—she's handy with a needle.

MacOrderly sighs. 'Ah right, boy. I have found your record now. Mr Arrol does this type of thing from time to time, although I wish to God he wouldn't. I have to balance the books. Training first. Where am I going to put you…?'

He consults another ledger bound in blue leather this

time, gets up and walks to the door, placing his fingers in his mouth. The ear-splitting whistle makes me wince. 'Harold,' MacOrderly hollers. 'Come here! I have a new lad for you!' He turns back to me. 'What's your name again?'

'John. John Nicol.'

The raven-haired Harold must be in his early twenties and wears a scuffed bow tie with his shirt and woollen jerkin. His eyebrows are permanently arched, making him look surprised or excited. From the start, I can't imagine him frowning. 'Welcome, John,' he smiles. 'I'll show you round the place first. Now, listen up, will you?' He strides off without looking back. I scamper after him.

The drawing loft and the sheds are the most enormous buildings I have ever seen. Harold leads me through the site first, pointing out entrances, exits, important people, delivery routes, downfalls and dangers—and too many processes to take in all at once. There are other boys working in here, although all of them look older than me. I stretch myself as high as I can and lift my chin to appear even a smidgeon taller than I am. From all angles and at every minute I feel eyes boring into me, assessing me, judging me. I may just be imagining that last bit. After all, these men and boys have actual jobs to do. Some men measure, others mark the huge metal pieces with chalk. Still others punch holes into the steel with intense precision.

Harold has clearly taught so many people to rivet that he delivers his well-practised lecture with the speed of a steam train, and all I can do is trust my brain to hold on for dear life. What he seems to be saying is that it'll be my job to heat the metal piece they call a rivet over a small stove until the metal is hot. An hour later my head swims with bridge words. The three diamond-shaped *cantilevers* are the towers which hold up the bridge, The *viaducts* are the arched towers carrying the train track onto the bridge, and the *lattice girders* will connect the cantilevers once the work is complete. He also says something about *strut* and *bay*, but at that point he loses me. Despite only practising the heating part, I burn myself twice in the process. Harold says that, when the rivet is hot enough, I will need to throw it—yes, *throw* it, to another man in my gang, the rivet catcher who will place and hold it in position before two senior riveters hammer it into place. Four men make up a rivet gang. Every time I leave the sheds, I see this very principle in action. It's as if all the humans crawling over the bridge like insects are just part of a giant, independent machine, following the same pattern, over and over again. *Heat, throw, hammer, heat, throw, hammer, heat, throw, hammer.*

My mind grows hazy with all the new impressions, the mix of molten metal, the smoke, sweat, drilling, cutting, shouting and scraping. But at least I have my feet on firm ground, and for the most part, a roof over my head as I work. *Nothing ventured, nothing gained*, as Dey would say. I am determined to try my hardest, and that is all there is to it.

6

Time and Tide Will Tarry for No Man

Crossing back to the northern shore at the end of the working day, every muscle in me aches. I am carried towards the train station in a river of briggers. Instead of making for the front, I hang back until the worst of the crowd has cleared and the new train pulls past me. It is then, in the half-light, that I spot something on the tracks behind it.

Waiting for the steam to clear, I look again. Yes, there is definitely something down there. Small, perhaps, and furry, though I can't be sure. I walk away, along the platform and towards the open doors of the train, but it is as if my conscience is attached to the thing by an elastic string, and I am pulled right back. Many of the briggers board the train and more arrive, but I stand glued to the spot, peering at the tracks. Going down there is dangerous, and absolutely forbidden—there are signs everywhere telling me that.

And yet, I can't help it—I need to know what it is. It

might be a purse, or something else that is valuable to someone, somewhere. I can hand it in. I should, in fact, hand it in. Resolutely, I swing myself down.

But it is not a purse. It is a squirrel, with the evening light dancing on its russet tones, even down here. Its soft fur feels wrong beside the hard iron tracks. It's motionless—it must have been hit as the train pulled in. I turn to clamber up to safety again, but something makes me look back. *Did it twitch, just there? I need a closer look.* With a deep breath, I pick the creature up into my hands. As I watch, there is an unmistakeable rise and fall in the tiny little chest. *It's alive!*

'Oi! What are you doing on the tracks!'

The voice comes from the other side, further along. I don't think about it, I just slide the injured animal into my jacket pocket and scramble back up onto the platform sprinting to the exit. The guard's shouts ring after me, but I am a fast runner and there are trees all around the North Queensferry brae.

Once out of sight and safely in the shadow of the woods behind the station, I stumble to a halt. 'There you go little one.' I tuck the dazed animal into my cap and push it into the undergrowth. That will give it some protection. On second thought, I also empty the crumbs from the greased paper that held my lunch into it. The steam whistle blows, and I just make it onto the train before it puffs towards Dunfermline and home. The wind tousles my hair on my way to our street.

'John!' Mother actually drops the washing she is hanging up by the fireplace and rushes towards me. 'How was your first day?'

'Very good, Mother,' I say without much enthusiasm.

By the time night falls, the fire in the range is dying down and mother and wee Helen are in their beds in the alcove. Dey is in his chamber off the hall. I am finally alone.

All of the images of the day sail through my mind. I am a brigger. Not a schoolboy anymore. I feel a painful pinch of longing for games in the playground and the bell that rings the beginning of the school day, for company of my own age. The moonlight casts the tree's branch patterns onto the plain wall beside the range. They remind me of the bridge, growing and jutting and angling this way and that. Perhaps, if I am very lucky, I shall be able to avoid going up on the bridge at all! All I must do is make myself indispensable to the workers in the sheds, so they won't let me go. I resolve on ways to achieve this. Bring in my mother's baking? Sing to keep people's spirits up? Become the hardest worker they have ever encountered? All of those things?

My lids droop as the last of the embers lose their glow in the grate. The moon is bright tonight, but its ghostly glimmer only speeds up my dreams. I close my eyes and breathe more deeply.

The next morning, I begin work in the sheds again. Soon I am sent to fetch tools from all over the place, in between running messages. I like running. Harold plays a trick on me by sending me to the tool shed for 'a long stand' and I am told to wait. It is half an hour before I realise the joke. The man at the tool shed desk laughs very hard at my expense, and so does Harold. But having a joke with the workers means that I am starting to belong.

In the evening, I cross the water with all the other Fifers, but decide to stop by the patch of undergrowth where I left the injured squirrel. Dey's cap may not be worth much, but it's the only one I have.

To my surprise, the squirrel is still there, curled tight into the cap. Slowly, I extend my hand towards it. *Is it dead?*

The answer comes in the form of two sharp front teeth, sunk right into my thumb.

'OWWW!' I bite my lip and shake my hand free. 'I rescued you!' I snap before realising I am talking aloud to a squirrel.

But there is no one here to hear me, so I say what I have to say. 'Fine! Go, and good riddance! That's my cap and I want it back.'

The squirrel looks up at me.

'Out! Get out of my cap!'

Perhaps it can't? Perhaps it's still a little stunned? If it was hit by the train, it can't have been hit all that hard— there is no blood. Absentmindedly, I fumble in my pocket and close my hand around the piece of bread I never ate at lunch.

I kneel down. 'There. Try that? Not your usual, I reckon, but…' The creature has begun to nibble the seeds off the crust. Carefully, I tilt Dey's cap until the squirrel uncurls its furry tail and lands on the ground, ears twitching.

'There you go. And now I have a train to catch. *Time and tide will tarry on no man*, you know.'

Clutching Dey's cap to my chest, I sprint for the train. I don't fancy the two-hour walk home from here if I miss it.

The next morning when I return, the squirrel is sitting by the roadside, as if it is waiting for me.

It crouches by the brae, cocking its graceful head sideways, its furry tail twitching to match its tufty ears. I stop. My hand creeps back into the pocket from which I fed it yesterday and yes, a few crumbs are mixed with fluff at the bottom of it. I bend to place them on the roadside and make to walk past. Behind me, a few workers *coo* and *aaah*, but continue to the boats as usual. The creature has devoured the meagre food and looks at me patiently again.

'Hey, what's your squirrel called?' laughs a passing brigger. I don't know why I answer, but the name comes out of my mouth before I can think at all. 'Rusty. But he's not mine. Not really.'

The days of the week pass in a similar fashion. I come on the early train and take the late one back—and yet, as soon as we briggers appear in front of the temporary site station at North Queensferry, the squirrel is there. I make a habit of keeping back a little of my supper for it, and it is surprisingly adventurous with its food. At first it does hobble a little, as if its front left paw was damaged in some way, but it improves until it is barely noticeable. It is a full fortnight before it will let me touch it. When I finally do, I wince on account of my burns from riveting training in the shed.

Before another week passes, I see a streak of rusty-red, shadowing me on my walk to the pier.

'Rusty, go home. I'll be back at night.'

But I can't help smiling as I board the boat, being pushed this way and that by briggers on all sides. *The little creature cares for me.*

My heart blazes like the red-hot coals in the rivet-stove.

7

A Good Tale Never Tires in the Telling

The following Saturday is the first day I am not required in the workshops. After the morning chores, I tuck the empty autograph book into my right pocket and *Treasure Island* into my left and set off on the short walk to the Carnegie Library.

I linger slightly less time by the lamp post outside, shaking my head to rid myself of the memory of poor David Clark. It seems only weeks ago that we sat together in the school room; he left school in the spring. I feel for his grandmother, all alone in the world now with no one to keep her company, and no breadwinner either. I send a speedy prayer to heaven for her. May she not be reduced to the Poorhouse at her age.

I take the entrance stairs in one leap and find myself inside. I can feel the cold seeping through my soles from the tiles on the floor. The air is still. It is strangely quiet in here—I expected it to be busier on a Saturday afternoon. The town and its shops are crowded enough. So much the better for me. I walk up to the library room where the librarian's boy is re-shelving a trolley of volumes.

'Is Mr Peebles here?' I ask after a moment's hesitation. The boy doesn't even look at me, but he answers: 'Mr Peebles is gone to his flat for luncheon. He won't be long now.' The youth must be fifteen or thereabouts; tall, dark-haired and awkward with a rash of spots around his chin. Just as I make my way back to the stairs to leave, the door flies open and the librarian staggers in under a pile of books so high that his face is completely obscured. I see it before he does, his foot hovers over an uneven floorboard. I rush towards him, just in time to catch the top five volumes as they slide off his tower. There is an undignified shuffle as we both struggle to recover our balance. His eyes crinkle into a smile as he recognises me.

'John Nicol? Well, well, so you're back. How did you like the Stevenson?' He walks towards the desk, and I follow him.

'I liked it very much indeed, sir.' Dey told me to *haud my wheesht* in the library, but I can't help it, my words tumble out in a waterfall of enthusiasm for adventure and night-time dealings, secrets and dangers. And Mr Peebles did ask! The librarian nods with interest, like one truly listening, not just waiting for a gap in the flow.

'Try this next then?' He ushers me back to the shelves, the same section where we stood before. 'Take a look, here. It's called *Kidnapped*, by the same author. It has not been published long. It's set in Scotland, not the South Seas. I have read it several times already, but they say *a good tale never tires in the telling.*'

That is exactly what Dey would say. I take it and carefully run my fingers over the bindings before giving it back to him. 'Thank you, Mr Peebles, sir.'

He stamps it for me as before, and I hold out my hand to receive the book, but he does not let go. I pull, but he holds fast. His smile widens. 'And John, did you remember the other thing we talked about?'

For a moment I am flummoxed. 'Oh, yes, my own empty book! I did bring it, sir, here it is. As I said, it's not much to look at, I'm afraid.'

He relinquishes *Kidnapped* into my hands and takes the empty autograph book from me, running his fingers through its crisp white pages. 'Tell me, do you know what to do with this, John?'

My neck suddenly feels very itchy. 'Famous people are to sign it, so my aunt says in her letter. But I do not know of anyone famous at all. And even if I did, how would I ever persuade them to sign their names in my book?'

'Well, young John, I took the liberty to look through some of my paperwork which I filed while establishing the collection here. May I?'

He places my book on the counter of his desk and

opens a drawer. With precise grips, he sets out a jar of glue which he unscrews swiftly. With the other hand he retrieves a rectangular piece of paper, cut sharply and about the size of my palm. He applies a little of the glue on the piece and presses it firmly onto the first empty page. My throat dries. 'But Mr Peebles!'

'Have no fear, my boy. Trust me. With this to begin your collection, you will soon be a proper autograph collector. Did you know that autograph books like this one can fetch a lot of money at the auction houses?' He hums very quietly under his breath as he makes sure the paper is stuck fast, and that no glue has spilled beyond the edges. Then he passes it to me.

I take it from him. It is the bottom part of a letter, for I can just make out the date and the lines of a letter paper. *Andrew Carnegie.*

'Andrew Carnegie? THE Andrew Carnegie?' I say incredulously—and much louder than I should have, for reminders of *'hush'*, *'haud yer wheesht'* and *'silence'* come from several ladies and gentlemen nearby.

The librarian leans forward, letting his monocle drop. He whispers. 'The very one. I have kept all of our correspondence regarding the establishment of this library, and this one was a note of little consequence. You are welcome to it.'

The librarian smiles again, then inclines his head towards the door. 'Until next time, John Nicol. Let me know of your progress. One piece of advice I would give you. Keep the book on your person at all times. One never knows when one will encounter a person of renown. One must be prepared.'

'I suppose so. Thank you, sir,' I mumble as I step through the double doors, through the tiled hall and back out into the street. There I look at my first autograph again. The *A* of Andrew is confidently swirled, as is the *C* and the *g* of Carnegie. The richest man in the world inscribed his name in his own hand and I, John Nicol, age 12, am the owner of it now.

The more I think about it, the less I can keep still. Before long I am sprinting back towards Netherton Broad Street and the house where Mother will have more chores for me—of that I have no doubt at all.

But I will do them gladly today—until the bell tolls after dinner and the low light of the sinking sun filters through the window: the time I secretly call book o'clock. Tonight, I'll begin to read *Kidnapped*.

8

It's An Ill Wind That Blows

Days merge into weeks, weeks into months. The early train chunters through autumnal forest patchworks, shot through with golden rays. The railway line glistens with frost and the flakes dance in the icy wind. The bridge structure deviously hides itself in the blizzards, or else the frozen fog rises from the Forth and settles on man and steel, making conditions so hazardous that more than once we run from the workshops to watch another unfortunate soul being carried towards the hospital on Hawes Hill, having slipped and fallen. By now, I do greet the other young boys working on the site, but make no mistake: we are not friends, and we don't talk much— there is no time for that. I am the fastest runner among the young ones, and more often than not, I am trusted with running the messages from one part of the site to another. Another young boy, Thomas Shannon, dies in a fall from the bridge—it is all the talk in the workshops for a day or two. I never met him, and I am glad of it.

On a couple of occasions, the storms are so wild that the men are called down from the bridge altogether. The hammering and the riveting ceases, but the drinking at the Hawes Inn begins in earnest then. I keep away from it all. Every break in the work I get, I pull out a library book from my bag and read a little more by the light of the glowing furnaces. The autograph book, on the other hand, remains firmly tucked into my jacket pocket. Try as I might, I am yet to spot a person of great renown around here. On the few occasions I have seen Mr Arrol or the two bridge designers, my hand may have travelled to my autograph book—but I lacked the courage to approach them. Still, best to be prepared. You never know.

It is at the beginning of March that Harold takes me aside at the end of my shift.

'John Nicol. A moment.'

I approach quickly, expecting another message or order. But judging by his face, I am summoned for another reason. Harold scratches his head awkwardly and looks past my face. 'John, you've proven yourself trustworthy and a careful worker. You're skilled with the rivets now.'

I see something in his eyes. *Regret? Fear?*

He clears his throat. 'It's time to join a gang. On the bridge.'

I can feel the blood draining from my face. 'But sir,

can't I stay in the workshops? Please? As you say, I have worked work well here, and I know how...'

'John, stop!' he interrupts raising an admonishing finger. 'It's the way it is. New boys are coming to the site all the time. They need to be trained here in the workshops just like you were, and you need to make room for them, like it or not. Here's the thing: everything we do is about the bridge. The *bridge*, John. Do you understand? Every single one of us is striving to complete this bridge. Look!' He gestures towards the looming giant rising from the water. 'Listen! She is calling to you, John. Hear the wind singing in the steel? That's the bridge saying: you're ready. Your place is up there now, with the others who can do the work as well as you can.'

My mouth begins to quiver. 'But I am a fast runner; you said so yesterday,' I protest weakly. 'That's no use up there!'

'You'll just have to learn to become a fast climber instead. Report to the site office first thing tomorrow.'

Like a criminal sentenced to death, I turn to slouch away, but I feel Harold's hand on my shoulder briefly once more: 'And John, I hear you'll work a gang with a Mr Murdoch and his two sons.' His voice drops to barely above a whisper, but with urgency. 'John, pay attention. You want to take care not to upset the Murdochs, d'ye hear? It's not a good idea to upset the Murdochs. They've a reputation. Do you understand what I'm saying? And now, off you go home, or else you'll miss the boat and your train on the other side.'

I look back twice. Both times, Harold is still standing by the door of the sheds, his gaze following me. When the boat pulls way, he raises his hand in farewell.

The next morning, as instructed, I arrive early at the supervisor's office.

My eyes flit to the huge bridge structure towering above us and my heart beats so hard it might jump right out of my chest. *Come on now, John!* Staring at my toes instead helps a little. A group of men draw near, and I try to smile as they walk past. MacOrderly calls them in. 'Ah, Mr Cain Murdoch, I remember you. Good. In you come, too, John Nicol.'

Wait. There are four men there already, but if I am to join their rivet gang, we should be four all together! I am confused and half-hopeful that they will send me back to the workshops. Cain Murdoch, the man at the front, doesn't even look at me as he strides past into MacOrderly's office. He has a face carved by the elements, a swarthy red beard and so many pockmarks around his eyes that I wonder whether he has held his face into the sparks of a furnace on purpose. I can smell a hint of ale off him, but his trousers hang loosely from his braces. The two stocky young men behind him look like twins, all muscle and sinew, with freckled faces and auburn hair. Their blue eyes slash into me like icicles as they walk by. The third boy is younger, with a sullen scowl. I scamper after them. What else is there to do?

'Morning. We were telt tae report here first thing?' Mr Murdoch does not sound like he really means to wish MacOrderly a good morning. He doesn't even wait for an answer from his superior. 'We brought my youngest son here tae make up the numbers of our gang.' The smallest boy looks like a gust of wind could snap him in two. He glares at the Site Manager.

MacOrderly is undaunted. 'Well now, Mr Murdoch. I wish you hadn't done that. Master Arrol himself recruited this rivet boy here, and now he's trained and ready and has been allocated to your gang. It was a very unfortunate business about your previous rivet boy and his severe burns.' MacOrderly's stare is as steely as the structure above us.

I gulp.

The Murdochs exchange glances. The oldest man speaks again, slightly louder, and it sounds as if he is barely able to control his anger. 'With respect, sir, we'd like tae keep it in the family; I telt you that when you gave us that last boy. Surely, I should choose who my wage packet depends on? Why should I take a chance on a bairn barely out o swaddling clothes? Look at him! And we dinnae know him!'

I retreat backwards, into the shadows of the office, but MacOrderly points to me, his voice even. 'Let me introduce you then. His name is John, John Nicol. Good. Now that you're acquainted, we can make some progress. Murdoch, you know very well that we gave your youngest boy a chance last year, and that we had to dismiss him for thieving. And imagine our shock when we realised that he didn't even go through our proper training! You really have a nerve to bring him here again. On behalf of Mr Arrol, I order you to take John Nicol up the bridge and be done with complaining before I tell the lot of you to pack your bags.'

It is only after MacOrderly's speech that I realise I have held my breath for all of it. The supervisor places his pen down on his desk, in perfect parallel with his ledger, and narrows his eyes at Mr Murdoch. The contrast between his calm control and Mr Murdoch's badly disguised fury could not be starker, but if the supervisor notices this, he hides it well.

Mr Murdoch's head spins round to me and his hand twitches. Flames rise in his eyes, but by the time he turns back to MacOrderly, he has regained control of his features. 'Very well, sir. If it has tae be that way. Boy, come with me.'

He addresses it to me, and I walk out of the office with some trepidation as Mr Murdoch clips his now redundant younger son around the ears, hard. 'Did you no hear what the supervisor said, you useless piece of dirt? Get out o my sight!'

The son barely flinches, even though it's clear he expected the blow. He just takes it, staring back at his father with the same soulless sullen eyes. Then the boy slinks away.

I may have often wished for a father, but Mr Murdoch? No. Rather no father at all than someone like him.

9

The Only Way is Up
March 1889

'Come on!' Mr Murdoch's words are shards of ice as he begins to mount the steps, with his two sons sneering back at me over their shoulders. 'Boy! What the devil are ye waitin for?' he adds. 'Up!'

First there is a sloping gangway. I try to slow my breathing, but I must also keep up with the Murdochs, so every third step is a jog to catch up. My rivet gang leader is a tall man, but his twins surpass him by half a head still, bulky and unpredictable. All their faces are dour.

'I've never been up the bridge till now, you see. At least it's not too windy today,' I say half-heartedly in attempt to make myself agreeable, buttoning up my Dey's wool jacket against the cold. No answer comes. We have reached the end of the gangway. In front of me, the Murdochs begin to climb one of the many ladders—*how do they know which one?* I don't have time to commit it to memory. Soon the first ladder disappears behind me towards the sea beneath. Each step sends my stomach somersaulting a little more.

We reach a higher, narrower platform. The only barrier between me and a fall into the Forth below is a makeshift railing made from a single rope. 'Mr Murdoch, please wait for me,' I squeak weakly, but he and his sons simply stride along towards another ladder as if there was no drop at all. I struggle to follow. All around me, men move up, down or along, manoeuvring sheets of metal, and grunting with the strain as they sidle past me. Far beneath us, waves crash onto the bridge's steel, sending tremors along the structure, and I force myself to look up instead. The clouds hang low today. That next ladder is long, like a never-ending set of rungs leading all the way to Heaven.

'Watch it, son!' snaps a moustached middle-aged man behind me and puts a hand on my shoulder. 'Not a good place to start swaying. Ye all right?'

'Yes,' I croak, though my ashen face is bound to tell the truth. 'After you.'

He swings around me and leaps up the ladder like a monkey. The Murdochs are almost out of sight! Clammy hand after clammy hand, I climb the ladder which moves creakingly with the wind. When I am almost halfway, the wood shakes, hard: someone else is climbing up below me. 'Hurry up, boy—I've a shift to get to!' In no time at all, the young man is by my heels and taps my legs in his impatience. 'Oi! Scurry!'

I want to be polite and acknowledge his complaint, but if I do glance down to see the tiny houses by the pier and the grey water and the gulls circling beneath my feet, I shall freeze and not be able to go either up or down. I strain every muscle and close my eyes, reaching for each rung blindly. This way I can pretend that scores of men haven't died falling from places just like this. Eventually, my hand grasps thin air. I have reached the top.

I am ashamed of it, but I lie flat on my stomach while my inconvenienced follower steps over me and runs— yes runs—along the narrow plank to yet more ladders. I pull myself into a crouch and reluctantly open my eyes. There is no sign of Mr Murdoch or his sons.

Where is my gang?

Suddenly, high above me, there is a roar. 'JOHN NICOL! WHEREVER YOU ARE, GET YERSELF UP HERE, OR SO HELP ME I WILL PUSH YOU OFF THIS DAMN BRIDGE MYSELF!'

A worker in a red cap along the platform gives me a sympathetic smile. 'You're with the Murdochs, are you? Keep your wits about you, that's all I'm saying.' With that, he returns about his business, twisting ropes together and moving tools and equipment. Miserably, I glance downwards. Men on the pier below look like ants. The waves froth in the distance and an invisible force inexplicably pulls me towards the edge. The man in the cap talks again, just loud enough for me to hear. 'Listen, there's nothing stopping you but your head. One step after the other, one hand at a time, lad. You keep your head up and then you go. We're all the same at the beginning.'

'Thank you,' I croak and crawl towards the final ladder, leading diagonally to the sky.

'JOHN NICOOOOOOL!'

Rage has become words. And it is my name.

Somehow, I manage up to the platform where the Murdochs are watching, their faces ablaze with hatred.

'Got the stove going for you,' a sweaty young Murdoch spits as he grabs me by the scruff of Dey's jacket and hurls me towards the glowing furnace. There is a makeshift shed erected around where we are working, giving the semblance of safety, but the thin wooden panels creak with every gust of air. I set to work, heating and throwing the rivets like I was taught.

From time to time, Mr Murdoch or one of his sons growl something at me. I can't always make out what they are saying, but try to act as polite as I can, with 'yes, sir' and 'aye, sir' and 'thank you, sir', even when Mr Murdoch deliberately throws the glowing pieces of metal back at me, claiming they are not hot enough yet. I know he is lying, but what can I do? Soon, my jacket and breeches bear the scorch marks of that man's sport while his sons laugh on. I have begun to be able to tell the boys apart—Canny Murdoch is the older and named after his father. He has a scar across his forehead where his brother George has none. Twice, tears would well up in my eyes at their treatment of me, and my fear would claim me, but I will not give these people the satisfaction. However skilled and polite and hardworking I am, they are determined to hate me. At school, these things counted for something—*being* friendly meant *having* friends; honesty and hard work earned the dominie's praise. I have never felt more alone.

Mr Murdoch keeps a tally of how many rivets we have driven in so we can get paid. I count in my head too. By the time the bell rings down below to signify the end of our shift, my hands and face are blackened by soot. I have two red-raw burns along my wrist. I need to take more care when removing the red-hot metal rivet from the stove—I touched the rim with my wrist by accident. That much is my own carelessness, but the marks on my clothes and the burn on my cheek is all down to Cain Murdoch, flinging red-hot metal back at me when I do not expect it. I can still hear their guffaws replaying in my mind—what sort of man takes pleasure in the suffering of others?

Still, it could be worse. I think longingly of the fireside at home, and that thought helps me to descend all the ladders, where often I have no option but to look down. At the beginning I simply close my eyes and feel my way down the rungs until I can bear it. At least the ground is coming closer with each ladder and slope I conquer. Other men and boys are going the same way and I try to blend in, almost losing sight of the Murdochs. Until I see that they are making their way back to MacOrderly's office.

Why? Aren't we all supposed to go home now?

I hesitate. The endless crowd of briggers splits into those catching the train to Leith and those who, like me, catch the boats to Fife and the other side. Against the shifting clouds, the bridge itself appears to sway and bend, motioning to the pier and MacOrderly's hut. A shaft of evening sun breaks through the clouds, filters through the steel and illuminates the path for me. Could the bridge be telling me to follow? It's madness. But my legs carry me after the Murdochs anyway.

Instead of walking to the main entrance of MacOrderly's hut, I creep round the back. It doesn't take long for the roar of Mr Murdoch's voice to shake the corrugated iron walls.

'I'm telling ye, it's no going tae work. With the boy I mean. We are family, the rest of us. That John Nicol is lazy, take it from me, and we cannae be doing with him!'

My stomach contracts. *What?*

'Aye,' one of the twins chips in. 'He was late tae the

platform for a start. And insolent—nae respect at all.'

MacOrderly hesitates. 'I have had no such reports about him from the workshops. Well, it's his first day. It takes a while to build up a head for heights, with some of these boys. As for insolence, he struck me as a likeable fellow, and it was Mr Arrol himself offered him the work.'

'The devil take Mr Arrol!' Mr Murdoch shouts before correcting himself, more quietly. 'I am simply saying that we only managed 380 rivets today.'

That's not true. I counted. We managed 481, and I heard Mr Murdoch claim to have done thirty more to the shift manager so we would get paid more. With every word I hear, bile rises in my throat at this man's dishonesty. His sons mumble something that sounds like 'aye' and 'that's right'.

Mr Murdoch speaks again. 'Sir, will ye no give my younger son a chance? He is strong and he is one of us. We'll work well together. He needs employment!'

MacOrderly must be standing up—I hear the chair scraping along the floor. 'You know very well, Mr Murdoch, we don't employ thieves and liars here. I will not have that boy on my payroll again, and neither will Mr Arrol. Listen, Murdoch, I have had to reprimand you for troublemaking in the past. Do not blot your copybook again. Your last rivet boy barely got away with his life after that burn, and I'm not convinced that you had nothing to do with that. You needed a rivet boy, and you have a rivet boy—and it is John Nicol, like it or not.'

There is silence as Mr Murdoch weighs up whether to protest or not.

After what feels like an eternity, the door of the office is thrown open and Mr Murdoch stomps out, kicking an empty bucket right across the pier until it nearly lands in the water. I dive behind a barrel just in time. He says another ungodly word and marches off, with George and Canny mumbling in his wake. 'Don't worry, Faither. We scared that last one off, didn't we? Just you wait.'

They hurry after Cain Murdoch who is walking fast and not turning at all, although I swear, I saw him nod.

I glance up at the bridge and bite my lip. There's an ill wind blowing.

10

Down and Down and Down

Mr Murdoch's pockmarked face follows me. It appears in the clouds, in the shadows on the path, lurks between the trees. It hovers in my mind as I feed Rusty the scraps from my luncheon and on the journey home on the crowded train. His voice echoes through my thoughts in the dark as I pull the blanket over me to sleep. I know not all men are good. Mother told me that, and I know it to be true, however much I would want to deny it.

The glow from the grate reminds me of my rivet-stove on the bridge, high in the clouds and never more than inches from death. Sparks in our fireplace speak of the red-hot rivets flying at me as soon as I turn away. In the half-light of our chamber, I wonder. *Can I do this—however much Mother, Dey and wee Helen rely on me? Do I have what it takes?*

Did I even sleep a wink at all?

I'm not sure. The sky is still dark, but it's time and I make my way to the outdoor cludgie to relieve myself.

Small flakes of snow swirl in the moody sky as I run back into the house and pull two layers of knitted ganseys over my head.

'John, your piece!' Mother shouts, exasperated.

Groaning, I hurry back to the door to retrieve the paper parcel containing a sandwich, an apple and a hard-boiled egg, jog along the road and sprint the last stretch to the station because I can already hear the steam whistle. On the train, my frozen fingers close around the rail. I must try not to think of the Murdochs. By now, a handful of the men in the carriage acknowledge me with a nod. I'm not one of them, not in the way that I belonged among the boys at school. I wonder if Jeremy is still first in the class. Who is second now that I'm gone? But the faces of my schoolfriends now give way to the stubble and sweat of briggers. I secretly study them. *Is this what my father looked like? Did Dey smoke and swear like these men when he was younger?*

I take a deep breath as the bridge slides into view through the dusty carriage window, part obscured by rain clouds. Then we pass a group of well-to-do houses high on the hill to the north and a sigh forms in my throat. I pat the autograph book in my breast pocket for luck. Then I do it again. I will need extra luck today. Perhaps today will be the day I will meet someone famous. Or even just someone notable.

But the Murdochs are never far from my thoughts, try

as I might to banish them. Perhaps they just had a bad day yesterday—after all, they had to bear a disappointment about their youngest and I mustn't judge lest I be judged. Perhaps today they will be civil, and tomorrow they may even be friendly. If I resolve to be my pleasantest self—will they bear my presence? I am my mother's son, people say. Pleasant, forceful, determined Janet Nicol— and her pleasant, cheerful son.

At North Queensferry, I emerge from the station where Rusty normally waits for me. The beginnings of spring shoots are appearing through the thin crust of snow, but there is frost in the air too. Where is the squirrel?

I almost jump out of my skin as Rusty lands on my shoulder, light as a feather. I force myself to stand still and enjoy the feeling of being trusted for a moment. He nuzzles into my ear before giving it a nibble. 'Get off!' I giggle and shake him from me, and he dashes into the bush, offended as I hurry towards the waiting boat.

There is no sign of the Murdochs when I arrive at Queensferry and reluctantly follow the steady stream of briggers towards the diamond-shaped cantilever. In the confusion of arms, legs, caps and tools, I am glad that I made a mental note of which ladders to follow. There is no other way. This is my path. Up. Resolutely, I swallow down the taste of panic, close my fist around the first rung and pull upwards. *Up ten rungs. Platform. Three deep breaths. Along to the second ladder on the left and up fifteen rungs. Turn right and walk along, south. Four deep breaths. Three more ladders up...*

I chant the combination in my mind as if it were a treasure map, like the one I read about in Mr Stevenson's book. Filling my mind with Jim and his South Sea adventures helps. The sun has begun to burn the top off the fog, causing the icy steel to drip. The rays warm my face and I close my eyes for a moment and breathe deeply again, trying to forget where I am.

'Oi, keep moving up there!' I am immediately scolded from beneath, but I cannot afford to freeze in fear here. My imagination is my only hope of success. I imagine I am a monkey from Queen Victoria's menagerie; or Rusty, jumping from branch to branch with skill and no fear. My knuckles are white, but I have done it: I have reached yesterday's platform. There is no one here yet, so I carefully tread along to the chest and fill the stove with coal, sheltering the flame with my hand as I light it. The ladders beneath me creak and a moment later, the little platform shed shakes with the steps of the Murdochs, scowling and cursing amongst themselves.

I have to try. 'Good morning, Mr Murdoch. Good morning, George and good morning, Canny.'

I do not look at them because that will only give them an opportunity to intimidate me. Instead, I keep my voice light, stoking the flames. Soon, the coals are red-hot and ready for the first rivet. I throw it in and reach for the tongs. Canny as the catcher crouches beneath the steel sheet we are attaching and holds the bucket ready, with his tongs poised. The other two pick up their hammers. Despite his age, Mr Murdoch's muscles strain beneath his shirt while I give an involuntary shiver and reflect on what a weakling I am. *Oh for goodness' sake, John Nicol, give yourself a shake! That is no way to talk! Dey always says that thing about how a misty morning may become a clear day.*

'First one coming up,' I say as I have been trained. When the metal rivet has turned the right shade of orange glow, I hurl it as precisely as I can at Canny Murdoch who catches it easily in his bucket. Speed is of the essence now. He inserts the hot rivet into the pre-drilled holes to join the two metal pieces together. Then he applies pressure to hold it in place, while George and Cain Murdoch take turns to strike the other side, woodpecker-fast, to flatten the rivet end until the pieces are securely joined. There is skill in it, make no mistake.

I have had the presence of mind to heat the second rivet already and throw it as soon as Canny gives me the signal. They mutter amongst themselves and shoot me dark looks from time to time, but I am resolved to do my work well, and nothing else.

By the time the gong sounds for a short break, my cheeks glow with the exercise and my wrist sports a new burn which smarts. I'll get used to it.

Mr Murdoch shoves past me to lean over the railing. I feel a little safer in the shed, especially now that the clouds have covered the sky and the wind is picking up again. Feeding the stove coal to keep it hot, I rummage in my jacket pocket for my sandwich when it happens. The autograph book falls from my pocket onto the wooden slats. For a moment, I panic that it may slide through the gaps, but something worse happens—the hand of George Murdoch snatches it up.

'Give me that! *GIVE me that!*' My high voice betrays my panic, and he knows it. He practically skips out of the shed. I am too outraged to feel fear now and spring after him. 'My book! My book!' George is a grown man, but he throws it lazily to Canny who catches it with one hand above his head. Mr Murdoch looks out over the water, but the sons stare at me defiantly, holding my book aloft over the thin wooden railing and daring the wind to tear its pages. I walk towards Canny, hesitating only when I remember where we are: intruders in the kingdom of the clouds, with nothing but shifting metal beneath us and the waves waiting below. But, as it turns out, anger is stronger than fear. 'Give me my book back.'

I try to speak with authority like the dominie, but authority is scarce when you are two heads shorter than your enemy. Canny chews his crust of bread with his mouth open. And this is where I make a mistake.

'I know you lied about how many rivets we managed yesterday, Mr Murdoch. I know that you are troublemakers and I have been warned off, but I can't help who I am put with. You *know* I am not lazy, so may God judge you for saying that. And now GIVE ME MY BOOK!'

Mr Murdoch turns, and I have never seen such hatred in a man's eyes, such evil. But he doesn't look at me. He looks at his boys. And nods.

Canny throws the book high into the air, away from the bridge, away from the platform and into the mist above the water below. I scream like an animal. But then I see that I have a much bigger problem.

Both the Murdoch boys are moving towards me, arms outstretched and muscles twitching, silent like wildcats in the night. The next thing I feel is a shove that takes me right off my feet. The wood of the thin safety barrier splinters apart and the world begins to tumble.

Down.

Down.

Down.

11

Where There is a Will, There is a Way

My arms and legs flail as I fall out of time, my face pulled into a grimace by my gathering speed as I plunge towards the water.

The icy air cuts through my clothing. My brain cannot process what it sees and hears: platform after platform swishes by, to startled shouts and hammer blows.

The pier is nothing but a blur. Death? I don't think about it. I think about nothing.

Until the impact.

The water's surface hits me like a solid steam train and knocks all the remaining breath out of me. It goes dark as the Forth swallows John Nicol. Salt tries to force its way into my mouth and into my lungs, and then, only then, do I realise that I am going to die. Today.

Mother. I begin to thrash with my arms. *Dey, and wee Helen.* I begin to kick my feet and wish I had been taught to swim. *Breadwinner.* That last word barely forms in my clouded mind, but it is enough. I try to hold my breath, to work out which way is up, to kick and splash, to force my eyes open.

There! I follow the light and kick; it is all I can do. But Dey's heavy woollen jacket drags me down like an anchor.

Where there is a will, there is a way. I kick, kick, kick, but the movement is slower each time as the cold claims me for its own. Desperately, I stretch for the surface just out of reach, but it is hopeless. Like a wind-up toy, my limbs stiffen and then refuse to move at all. The darkness of the deep envelops me. There is nothing more to do but to close my eyes. *God have mercy.*

Just at that moment there is a heavy splash as a shadow moves over me, blocking out the last of the light. I feel movement in the water beside me. A small hand grabs the fabric of my jacket and pulls it up, and me with it. Thin legs kick hard next to me. Finally, I can't hold my breath any longer and open my mouth to breathe, but salty water fills it and I allow it.

A moment later, we break the surface and the cold air slices into my soaked skin all over again. More hands drag me from the water and heave me into a boat. I lie sprawled on the deck, coughing and retching, before my stomach empties itself of the sea within it. There is a thud and a crumple and a crush as another body lands beside me. My rescuer has been hauled from the water too. I barely lift my chin to see a mess of seaweed and petticoats—a dishevelled mane of dark curls.

The one who jumped into the water after me...

Was a girl.

I wake as a man wrestles my jacket off me and replaces it with a blanket, concern in his eyes.

'Careful with that. It's my Dey's,' I slur.

'Good. You can speak—sort of,' the man smiles, but I can still see the traces of terror on his face. 'You better be worth it—my Cora risked her life for you.' He nods sideways where a skinny dark-haired girl shivers under a blanket just like me. 'We'll get you both to the site hospital. It won't be long now. Name?'

I grit my teeth in the effort to speak clearly. 'John Nicol.'

The man takes a note in some sort of log paper on a clipboard. His boat is like a floating version of a lost property office, with caps, jackets and tools piled up in the corner. *Wait, is that...*

'My book!' I can barely raise my arm to point, but it is unmistakeable: there, still in the net, is my autograph book.

'That's yours, is it? We were just fishing it out when you followed. I can't believe you survived that fall!'

He strolls over to pass me the book, dripping wet on the outside. I carefully open it—it's a miracle! The pages are almost unaffected inside, only wet around the edges.

I stare at the precious Carnegie signature. Only now I realise what nearly happened to me, and my eyes fill with hot tears.

'Thank you,' I croak to the man. 'Mr...'

'Ramage. I'm the captain of this rescue boat. But don't thank me. I didn't pull you out, though I saw you go under. We threw a rope, but you didn't even try to reach it. And when we lost sight of you, we thought you were, you know...'

I don't respond.

The man shakes his head. 'Well, the next thing I see is my Cora, only twelve years of age, throwing herself into the water dress and all, instead of giving you up. I'm telling you, that's not what I had in mind when I let her come today. Thank her. Not me.'

I turn to look at the girl again. She is shivering violently with her teeth chattering hard. 'No need,' she gets out before I can form a sentence.

Mr Ramage's deck hand steadies me as I make my way onto the Hawes Pier, past the inn and to the hospital by the workshop. It is all so familiar, and also not at all: I have never been inside before. I follow Mr Ramage who carries his daughter Cora in a shaking bundle. As soon as we enter, nurses swarm out like bees. I allow it all: stripping me of my wet clothing, washing, combing, poking and prodding.

The doctor comes—I have seen him before. He asks

questions, moves all my limbs and checks my neck and my eyes. A reporter from the newspaper stands beside and scribbles on a notepad. Finally, MacOrderly arrives and thanks the doctor before turning his attention to me.

'John, thank God, you're all right. I've sent word to your mother already. I'm told the doctor wants to keep you in for a night's observation, and the same with the girl. Now, I just need to file a report.' His pencil hovers over his incident ledger. 'As best as you can remember, John. How did you fall?'

12

The Truth Will Out

The next day Mr Ramage and Cora appear at my bedside, just as I gather my bundle of tattered possessions. Even the edges of my precious autograph book have been dried by the hospital fires and I clutch it to my chest gratefully. We both survived, my book and me, and waves of relief wash over me whenever I think of it. But a new nagging fear has taken hold of my heart. *What will happen now? Was I right to tell the truth about what the Murdochs did?* MacOrderly's face darkened as I spoke. He scrawled an aggressive note into his ledger, turned on his heels and walked out like a man who had things to do. Difficult things. Necessary things.

Now the Ramages guide me towards the Hawes Inn. They've invited me to celebrate with them after our release from hospital, and it would have been rude to refuse, wouldn't it? Besides, I am nervous about returning home. What will I say to Mother and Dey? If the Murdochs are to be sacked, I will have no gang.

And perhaps no job? What will Mother say then? Just another thing to worry about.

'Do you think I was right to tell the truth? About my rivet gang?' I ask Cora who has occupied one of the tables, near the fire. There is bound to be a chill in her still, just as it is in me too. Her father leaves to speak to the barman and for no reason at all, I confide in her with everything.

'Of course, you had to tell the truth, John. Attacking you like that was a terribly wicked thing to do, whether they really meant for you to fall or not! And didn't you say that the boy they worked with before also came off worse? No, John, I understand that you didn't want to complain about them, but you had to tell the truth about this. Imagine how I felt when I saw you crash into the water right beside out boat! The sound! The splash! I didn't particularly fancy that water, but—'

'I'm grateful.' I say it quietly. I hate being in anyone's debt, and I hardly know her. I can't believe I have spilled my sob-story about the Murdochs to her.

'Well, you can pay me back by being my friend when I'm here at Queensferry with Papa.' She says it in a way that makes me wonder whether she is jesting or not.

Suddenly, Cora gives a very sharp intake of breath beside me.

'What?' I whisper.

'Look over there. Mrs Margaret Moir!'

I stare across the room. The lady looks remarkably elegant and confident, like any other rich and privileged lady I walk past in the street. She is surrounded by a mingle of men in suits and top hats. 'And what about her?'

Cora looks positively shocked. 'John, she is married to a leading engineer—her husband is in charge of a whole section of the bridge! Haven't you heard of her? She describes herself as "an engineer by marriage", and she did some great work to make the foundations safer for the workers at the time, and she…' Cora runs out of breath. This gives me the chance to interject. 'Breathe! Goodness. Why do you care so much about her? There must be hundreds of engineers working here.'

Cora has gone quiet. 'Because that is what I want, John. To be an engineer. I love the thought of building and creating, and I love this bridge. It's the truth, so don't look at me this way. I pester Papa to take me whenever I am not needed at home.'

There is silence.

'I am going to be an engineer.' She repeats it so quietly that I can hardly hear it.

'So, why don't you go and speak to her?'

She shakes her head bitterly. 'What would I say? "Excuse me Miss, how does the daughter of the skipper of a rescue boat train to be an engineer?" Don't be ridiculous, John.'

Just at that moment, the lady and her companions leave by the exit furthest away from us and Cora's father returns with two steaming bowls of porridge. 'There you go, something to warm you up before you go home. Cora, you and I have to put in another shift on the boat today I'm afraid. Your mother sends her love. We have half an hour before we need to start, so eat up.'

But his daughter simply stares wistfully at the door through which Mrs Margaret Moir left.

Her father seems to be used to her daydreaming, and shrugs, turning his attention to me instead. I stir the porridge and lift a steaming spoonful to my lips. 'Thank you, sir,' I remember to say, just before relishing the warmth of the liquid down my throat. Mr Ramage leans over so close that I can see each individual hair on his tangled beard. 'John! See that man by the window?' He flicks his pinkie in a barely noticeable gesture.

The person he indicates is flamboyantly dressed and barely more than skin and bones, with neatly parted fine, roguishly long hair disappearing into his collar and an impressive moustache twisted upwards in elegant curls. His eyes gaze intensely at the bridge, then back to his notes on the table, then at the bridge again. His fingers curl around a well-worn pencil. It is late morning, and I can't tell what he is drinking. He is alone. Beside me, Cora begins to eat too.

Mr Ramage leans in closely and whispers: 'The

barman tells me that's the famous writer man. The one who writes books and serials for the papers and all. Stevenson, they call him. Robert Leo Stevenson or some such. He likes to visit, though he travels much these days.'

My spoon sinks back into its porridge bowl. '*Louis*. Robert *Louis* Stevenson! Are you saying that's him?' I have said it aloud, and much, much louder than was wise. The gentleman has lifted his head, looking around to see who has said his name.

And I am on my feet and walking over, straightening my hair with one hand, and pulling my autograph book from its pocket with the other.

13

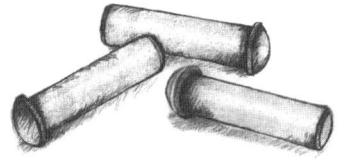

To Travel Hopefully

I cannot believe my luck! The author was so obliging, if a little preoccupied. He beamed with pleasure when I told him how much I had loved *Treasure Island* and *Kidnapped* and signed gladly on the page following Andrew Carnegie. Two famous signatures and a brief bath in the Forth—what adventures my little book will have to tell!

Cora and her father have left for their boat, so I finish my porridge in a happy reverie of these unexpected blessings.

When I head for the door of the Hawes Inn, however, there is a commotion outside. I have barely let go of the door handle when I nearly bump into the barman, blocking the entrance with hands on hips. In the street beyond him stand three men.

I'd know those voices anywhere.

'...it's a free country! I can enter an inn and drink if I want!'

'Not anymore—I banned the lot of you after the last time when you started a fight. AND I know for a fact that you have all been let go for what happened to that boy. It's the talk of the site!'

Somehow their eyes come to rest on me, standing in the open doorway.

Cold fury is carved all over Mr Murdoch's face and both sons have their fists balled, with blood vessels protruding. The barman points to the road. 'Go! I don't ever want to see any of your faces again. I'm sure neither does Mr Arrol. There may not be enough evidence to have you locked up, but the Site Manager believes the lad, and after the way you have conducted yourselves, I am inclined to side with him. Over here! Give me a hand here, everyone!' The barman gestures over to the site security guards which, finally, has the desired effect—the Murdochs slowly shuffle away towards the road, making gestures and cursing. 'This isnae the end of this!' shouts Mr Murdoch. 'We'll pay ye back. Mr Arrol, you—and especially that liar, John Nicol. I swear on my mother's life.'

The security men have reached us.

'That's enough, Murdoch! Move on!'

Mr Murdoch turns and walks up the road now without looking back. But his sons look straight at me. And Canny draws his hand across his throat before spitting on the ground.

On my return home, Mother enfolds me in a close hug, the sort of embrace she hasn't given me for a long time. Dey ruffles my hair a little longer than usual and Helen clings to me like hoof glue, stretching to reach my waist and lifting her arms for me to pick her up.

'John, I'm not sure I can bear you returning to the bridge at all. I can't risk it,' Mother whispers that evening, looking intently at the pot of stew she is stirring on the range. Her voice is higher than usual.

I shrug. 'I've got it out of the way now, don't you see, Mother?' I say as light-heartedly as I can. 'Nothing left to be afraid of.' I lean over her shoulder and snatch a spoon for a quick taste.

'Nothing but falling again, and not being so lucky,' she mutters through clenched teeth. I put my arm around her shoulder, even though I am still a little shorter than her—but I don't hold her for too long. Too long and she may cry, and I can't be doing with that. Dey mumbles something in the corner in reply, but I didn't catch it and I don't ask.

The next morning, I emerge from the station and wander down to the pier. As I walk, I keep my eyes open for Rusty, but no rustling in the bushes, no swinging twigs, no half-eaten cones. Perhaps, like me, my little friend has embraced making his own way in the world. I shouldn't be sad or disappointed—but perhaps I am, a very little.

I report back to MacOrderly's office, standing an inch taller and wearing my Dey's cap at a jauntier angle than before. I have a new gang. The Murdochs are gone, and I have been given five shillings of injury money too, as well as a personal message from Mr Arrol, and Mr Baker and Mr Fowler, the designers. It's a shame that it wasn't a written message—I could have pasted the note into my autograph book!

'We're sending you back up the bridge, John. Mind your step—but O'Malley and his gang are decent men. We owe you that now. They call him Bristles, on account of his beard, I think.'

'Will the Murdochs be back, sir?' I ask tentatively.

'I certainly won't employ them again, but I can't stop them from coming to the town, unfortunately. I suspect that Cain Murdoch will walk into this office most mornings to give me a mouthful for a while, but don't worry yourself, boy. They will soon get the message. We are not for turning, and, in time, they will pack their bags and be gone.' He hurriedly shakes my hand before shouting 'NEXT!' and waving another worker in.

It takes me a moment or two to calculate the platform which I am to report to, but a minute later I am on the walkway and five minutes later the pier and the waters below disappear beneath me once more. Up, up and up again. Of course, my stomach still somersaults, but if I look up and breathe deep despite my fear, I can control it. This realisation makes me so giddy that I almost skip onto the final platform where the freckled old Irishman nicknamed Bristles O'Malley and his gang are waiting for me. The stove is already hot. Without hesitation, I put down the clinking bag of rivets from my back and place the first into the glowing coals with a smile. My first throw is so accurate that the sturdy rivet catcher Doug raises his eyebrows with a 'well done' and I feel I could soar with the gulls, so happy is my heart beneath the autograph book in my breast pocket.

I travel hopefully. Things are looking up.

14

A Penny Wise and a Pound Foolish

Cora and I sit on the pier below the Hawes Inn one April Saturday evening. I've put in my shift already, and I am making a list of dream signatures while Cora talks.

'So, top of your list would be Queen Victoria's signature, but she is getting on in years, is she not? I am not sure she'll sign something just for you, unless you do something terribly brave in a war. So perhaps you might settle for the Prince of Wales. He'll be king next.'

I laugh at her fanciful talk. 'Prince Edward? Please, Cora, let's aim for people I may have a chance of actually meeting.'

Undeterred, she carries on. 'Alexander Graham Bell, the scientist and inventor, perhaps. I long to see a telephone—can you imagine speaking to someone who is not standing beside you? It's simply mind-blowing! And there is a writer called James Barrie, he is a Scot, I think. Of course, there are all manner of engineers—like that Frenchman Mr Eiffel—have you heard of the tower he is building in Paris? Perhaps he is finished, I don't know. Mr Darwin is dead now, isn't he? But perhaps Lewis Carrol, he wrote that charming story of a girl in Wonderland. My father read it to me. And all sorts of leaders of countries, like the American President and—'

At this I cast my pencil onto the pier and double over in laughter. 'You're mad, I tell you. But perhaps, if I have very good luck, I will one day run into Mr Arrol again, and I pray I'll have the courage to ask him then.'

Cora shakes her brown curls. She is in the habit of doing away with her bonnet whenever she comes to the site. 'If this autograph book is to be worth anything to your family, you must maintain the standard. Mr Andrew Carnegie and Mr Stevenson, these are big names, John. As big as they come around here. Have faith!'

I reach for the pencil again to resume my thoughts, but something causes both of us to turn around. There is some sort of din. Ever since the railway workers arrived to construct the approach railways for the bridge, the town has been a rowdier place than before, and that is saying something with all the briggers in one place. But this is different. The streets are filling with men, one as drunk as the other.

'We should go,' I say to Cora, gathering up my book and pencil.

'Go where, John? I am to wait here until Papa finishes his shift on the boat, that's another hour.' I point at the gathering crowd. 'Fine,' she sighs, and we rise. From one minute to the other, fury fills the air. It's definitely time to go.

We crouch low and make our way further up the pier, ducking behind a pile of boxes to stay out of sight. Every now and then, I glance nervously at the fifty or so men now hovering outside the inn.

'What do you mean no?' one shouts at the poor doorman. There is shoving. I hear glass smashing. A lady screams. Much roaring and shuffling ensues.

'Don't let anyone tell you what to do!' shouts a dreadfully familiar voice, and I can't help taking another peek. I should have known. Right at the front of the mob, waving a long rod of steel threateningly above his head, is Mr Cain Murdoch. Foamy flecks of spit fly from his mouth: 'Boys, let's give these masters a piece o our mind, right? They hire us and fire us as the whim takes em, with nobody tae speak fur us. Do this, dinnae do that. I say: let's no stand fur it any longer! Power tae the workers! And tae our right to have a dram whenever we want one!'

There are roars of support—these people are a penny wise and a pound foolish! Can't they see Murdoch for what he is? I watch just long enough to see Mr Murdoch bring his piece of metal down into the window of the inn. The smash is quickly obscured by the shouts from the crowd. There must be hundreds of men on the streets now. Some of them force their way into the building and soon emerge with several casks which they noisily open and pass around, pouring what looks and smells like whisky straight into their mouths. Cora claws into my arm.

'We need to get help!'

The mob has swelled further, and Mr Murdoch has climbed onto the shoulders of his two sons, shouting loudly from his elevated position. 'Listen! Mr Arrol will try tae take all the credit, but it is US who built that bridge, all of us together. And now he's given me an' my boys the sack for a silly wee accident.'

It is my turn to hold Cora back who nearly jumps out from our hiding place to give the liar a piece of her mind, but the rowdy applause would have drowned her out in any case. The first of the police officers arrive on horseback.

The noise is deafening now, so Cora has to lean in close and speak right into my ear. 'There aren't enough policemen! Poor officers, there are just so many of these rioters.'

She is right. Canny and George Murdoch have helped their father back to the ground, but like some of the men around them, they have picked up stones from the ground and start hurling these at the policemen. One of the police horses shies and rears.

'That's it!' shouts Cora beside me and before I can stop

her, she sprints towards the mob now oozing from every hut on the hillside, shrieking. 'Not the horses! Don't hurt the horses! Stop! STOP!'

I have no choice but to follow her. She'll be trampled or hit by a rock—or another disaster I can't even imagine yet will surely befall her. She saved my life, and I cannot abandon her now. I delve into the crowd after Cora, crouching so that my face won't be visible. But within moments, I have lost sight of my friend and the green pattern of her dress.

15

On the Safe Side

I had hoped to catch the last brigger train back to Dunfermline, but it looks like I'll have to walk tonight, and certainly not before I have made sure Cora is safe. Everybody around me is shouting. What can be the harm in calling for her?

'CORA!'

I catch a brief glimpse of my friend, pushing back men at the front of the mob, but she is elbowed out of the way and crumples to the ground before crawling aside, inches from the heavy hooves of the police horses. 'CORA!' I shout again but as I try to make after her, the scruff of my jacket yanks me back.

'John Nicol!' slurs the drunken voice of Cain Murdoch, his blood-flecked eyes boring into me. 'I've been looking fur ye. There is a score tae settle between us, John, and I plan tae settle it now.'

He raises the metal rod he is still holding, and I duck down. Instead of pulling away, I ram my head into his stomach as hard as I can, which does the trick—he is momentarily unbalanced and the grip on my jacket loosens. I dive to the ground, hearing a loud curse as his weapon clatters onto the cobbles and his two hands meet in the air where my head was. I do not wait to see what else he has to say, or for his sons to get involved. I crawl through countless legs, away, away, scraping my knees on the ground. Only when I deem myself beyond their reach, I jump to my feet. There—a fleeting impression—Cora's dress disappears around a corner by the Town Baillie's house.

I run after her as fast as I can, but it takes me several minutes to find my way through the mob. I arrive just in time to help her hold the reins of the Baillie's horse as he mounts it with his truncheon, and he sets off.

'The Baillie and his officers are going to do what they can to help. What is wrong with these briggers?' she shouts, hoarse with the effort.

The Baillie canters downhill in the direction of the disturbance—soon the sounds are a mix of police whistles, horses whinnying, shouts and truncheon blows.

It takes another hour for the street to be cleared. A number of men stand handcuffed outside the Hawes Inn, but—my stomach clenches—the Murdochs are not among them.

'Cora, where have you been?' Mr Ramage embraces

his daughter. 'Running errands for the Baillie,' she says, deadpan, her tugs of hair outlined by the pale moon rising behind the bridge.

I sigh. 'I have missed the last train home!'

It is true, and Cora and Mr Ramage look concerned. 'How are you even going to get to the north shore of the water tonight?'

I hadn't given any thought to all of this. The thought of staying on the streets of Queensferry with the Murdochs on the loose tightens my throat. I say nothing.

Cora looks at her father who exhales slowly before nodding. 'I suppose we could take you across before we go.'

'Thank you, Mr Ramage, thank you!'

The seaman man turns in the very direction from whence he came and unties his boat. I help him throw the ropes aboard while Cora powers up the engine. The fog hangs low over the dark waters of the Firth, and as we glide across the river, the sounds from the shore fade with all their troubles. Now all we hear are the splash of water against our boat's wood, while the ghostly steel-structure of the bridge rises from the mist in the dim moonlight. The lattice girders cast beautiful shadows onto the water's surface as above us, almost forty thousand tons of steel sway and creak. During the day, I don't notice it amid the shouts and the hammer blows, but tonight the bridge is alive—I have no doubt of it.

Cora lies half asleep on the bench by the side of the boat, but her father holds my gaze before looking up towards the sky through the bridge's pattern. He hears it speak too, just as I do.

Once landed on the northern side, I say my goodbyes and thank the Ramages, before setting out on my two-hour walk home. At least there is a good body of water between myself and those who bear me a grudge, and that's all that matters for now.

I set out, dragging my weary feet along the long road home.

At one point, I think I see a rusty-red streak along a branch above, but I've probably imagined it.

'What in the world of the Lord,' begins Mother when I squeeze through the front door at home before dropping her voice to a hissing whisper. Helen must be asleep already. And I can hear Dey's snores from the small bedroom too.

'I expect you'll hear all about it at church tomorrow, Mother; there was rioting, briggers and railway workers too. I am spent.'

Her face softens and her hand ruffles through my hair. 'There is some dinner left on the range. It'll be dried up, mind.'

Now that she mentions food, I realise that I could probably eat a horse, so hungry am I. 'I'm sure it'll be good, Mother. Thank you.' I tiptoe into the kitchen and scrape the remainder of the fish stew from the pot as quietly as I can. Mother startles me as she appears in the kitchen again, her shawl wrapped tight around her nightdress.

'I am glad you're home, son.' Her eyes glisten. 'When you didn't return on the train, I thought, perhaps…'

She does not finish the sentence. She doesn't need to.

'Don't worry about me.' I step over to hug her tight and hear a tiny sob of relief. 'I'm a breadwinner. Nothing will happen to me. And Mother—am I taller than you?'

16

Enough is as Good as a Feast

Spring turns into summer and the progress on the bridge is astonishing, even if the famous designer William Morris describes it as *'the supremist specimen of ugliness'*—his remarks are the talk of the site for a day or two. The cantilevers are nearly joined and soon trains will be travelling across. It won't be long. From my vantage point high on the bridge, I can now look down without shuddering. I watch as the hillsides around us begin to blossom with gorse. My eyes roam from Arthur's Seat to the Pentland Hills in the south, along the sandy beaches hemming the mirror of the Firth of Forth, and across to the distant horizon where water meets light in a blurry haze. Below us, pleasure cruisers weave around steel and rock, full of people who want to see the bridge up close. Then something catches my eye. Right beneath us some of the workers are constructing a platform.

'What's that for? I ask the rivet catcher Doug, who has come up beside me unwrapping his piece. He has pork and egg in his sandwich today, I notice with a little envy.

'Some fancy visitor. I spoke to Elizabeth in the offices this morning and she thinks it must be someone quite important. Even she hasn't been told anything more yet, other than it's happening in July.' Doug's eyes shine and I am certain that Elizabeth from the offices occupies a great deal of his thinking. But I also feel a shiver of excitement travel down my spine.

Doug looks puzzled. 'What is it? Why are you smiling, John?'

I look at my young colleague uncertainly. 'I have this autograph book, you see...'

To my surprise, he doesn't laugh. In fact, quite the opposite is true: he asks if he could see my book and then he calls over Old Bristles O'Malley and shows him too. I feel a little easier.

The Irishman winks. 'Sure, be leaving it with me now, John. I'll find out, so I will.'

The following day he calls me aside. 'Come here, John, I'm only after asking around, and the word on the street is: we'll be hosting the Shah of Persia.'

'The who? Of what?' My head swims.

O'Malley shrugs. 'He's the King of Persia, I'm told. Or something like that. Fierce important and exotic gentleman from the Commonwealth, now.'

Royalty! The Shah sounds just like the sort of gentleman whose signature should appear in my book, but how on earth I am going to get near him on the day is a mystery to me. *Think, John, think.*

'Thank you very much, Mr O'Malley. Thank you. Oh, holy smoke!'

I have seen something. No, I am not mistaken. There is a squirrel on the bridge—on the actual bridge! I can see it coming closer, just past my gangmaster's bristly face. 'See that there, Mr O'Malley?'

'Faith and begorrah, it's a blasted squirrel, so it is. How did that get up here?'

Doug has heard us and despite the slight swaying that the bridge does every windy day, he throws the creature a crumb. We wait patiently. For a moment, the squirrel has frozen stock-still, like a furry statue. All around us, there is the daily deafening drumbeat of metal on metal, but it doesn't move. I can barely believe it, but it's him, all the way out here. He must have run along the ropes connecting the cantilevers as the metal connections have yet to be inserted. The animal turns its head to the side.

'Rusty?'

I say it loud enough for the men to hear.

'You've named the thing?' asks Bristles O'Malley, but I only have eyes for the animal. 'This is Rusty. I feed him most mornings in North Queensferry. Months ago, I discovered him hurt on the railway tracks. Look, this one ear has a chunk missing, but not the other. Rusty!'

To my delight, both ears tilt forward, and the squirrel

sniffs the air, as if remembering. In short jerky jumps, it comes towards me. Doug throws another crumb. '*Rusty,*' I purr trying to imitate the voice I used when I first found him all that time ago. Finally, the animal reaches my hand and takes a bit of my piece from it.

'Told you, it's him,' I say proudly, but the squirrel has already retreated again. In elegant, secure arches, it bounds from plank to plank, downwards and across.

'Poor thing. Do you think it was lost?' Doug asks.

O'Malley shakes his head. 'Nah. I reckon it came here to find John, so it did. But back to work, boys—unless you're content to be doing without pay this week.'

Back on our platform, I stoke the coals into sparks and add another layer before placing the rivet in it and checking that Doug has his protective leather apron on again before I throw it. *Just wait till I tell Cora. The Shah of Persia indeed.* I get a strange pleasure out of simply saying his name. *Shah of Persia, Shah of Persia, Shah of Persia.* Even the name spells adventure, like an adventure novel from the library.

***

'Do you have anything on the Shah of Persia, Mr Peebles?'

The librarian looks at me, momentarily flummoxed. 'Why are you interested in the Shah of Persia, John?'

I wave him forward and he leans on his desk until there can't be more than a hand's breath between our noses. 'Because he is coming to visit the bridge,' I whisper.

Mr Peebles pulls back in astonishment. 'Is he? Is he indeed?'

'Yes. And I am going to need your help—how do you think I can get his autograph for my book?'

The librarian scratches his head, then his beard, then his head again—it is something he does. 'Well, we do have several large volumes on Persia. Let me have a look.'

I envy the librarian. He alone is allowed to walk wherever he wishes between the shelves, he may arrange and restock books, ascend the ladders, and treat every one of these books as his own. In the distance, his assistant carries a pile of large books to the smoking room, staring ahead, his mouth half open. His position may not be as well-paid as a brigger's, but he is making a wage without risking death. I have tried to speak with him, but I sense nothing from him, no love, no spark, no imagination. He wheels the trolley of books from one end of the library to the other, slams the beautiful volumes in their places as if they were sausages, or bricks, or pots. The boy doesn't seem to understand how much they matter, these books. I once saw him tear a page in carelessness as he replaced a slim book on a high shelf. He looked around, but Mr Peebles was not there to see it, so the boy swiftly hid what he had done. I almost felt the book's pain, but I said nothing. What is there to say?

I have seen it: whenever Mr Peebles himself is unoccupied, he begins to leaf respectfully through some tome or another, reading, understanding and learning more.

Mr Peebles is back, bringing a book and a newspaper. 'What you say appears to be right, John. It does look like the Shah will not only visit London but make the trip to Scotland too. It says so in the newspaper so it must be true.' He chuckles at that, and I am not sure why. 'As far as books are concerned, all I can find is this: *Travels in Persia, 1673-1677* by Sir John Chardin.' It's not exactly modern, but it may give you a flavour.'

'Thank you, sir. But Mr Peebles, how am I to get the Shah's signature? Can you think of a way? You are the cleverest man I know.'

At this he laughs properly, though everything about the man happens quietly. 'I do not have the first idea, I'm afraid. I think this particular scalp may be out of your reach, my boy. If I think of a genius idea before his arrival, I'll let you know.'

I sigh. 'It's almost here. The visit, I mean. We have been told that no one will carry on working. All of us have to stop so that the Shah can take a tour. There will be all manner of important ladies and gentlemen. I wish I was more than a brigger!'

The last sentence came out slightly louder than I had anticipated. A few people in the reading room turn their heads before sinking back into their plush chairs. Mr Peebles, however, gives me a very stern look. 'John!

Never speak so again. You *are* more than a brigger. You are a reader, a man in the making, with thoughtfulness and imagination and spirit. You are enough, and enough is as good as a feast. Don't ever let anyone tell you otherwise.'

He slams the book onto his desk, stamps it with much more gusto than he normally does, places it in my hand and turns to the lady waiting for his attention by the window.

I am dismissed.

17

Hurrah for the Shah

The big day has arrived. It makes a change from my usual morning shift—instead of climbing to the heights, our gang, and many like us, are given the task of assembling the decorations for the Shah's visit. The Irishman Bristles O'Malley nearly spits when he is handed a neat pile of steam-ironed Union Jacks to suspend from our cantilever, but the lady in charge is not to be dissuaded.

I am surprised at myself—it has been months now since I felt that paralysing fear up here; from a distance people must think that I am moving across the high platforms like circus acrobat. I am certainly more agile than the rest of my team who pass me pieces of string and hand me the next flag, and the next.

That done, we become part of the effort to drape crimson and gold cloth along the path onto the bridge. A huge, ornately lettered *'Welcome'* sign already hangs from the arch above. O'Malley mutters unhappily under his breath as we finally help roll out the crimson carpet across the jetty. 'Stuff and nonsense!' he huffs, louder. 'I'd rather be building the bleedin' bridge than be tying ribbons for royals, so I would. Blasted Commonwealth!'

I ignore him. It's hard not to be swept away in the swirl of excitement. It feels as if I have walked right into the adventure stories I love to read: my two worlds are colliding. Gradually, the special grandstand beside the reception platform fills with well-to-do spectators. Every street in Queensferry is packed with onlookers not able or willing to pay for a ticket, and the sharp hoofbeats of the specially assembled police officers on their horses can be heard patrolling the town.

Finally, O'Malley, Doug and Jim Smith line up with me behind the barricades on the jetty. It feels like every brigger is here now, shoving and gossiping, smoking and stretching. When will the important visitors arrive?

'John! There you are!' Cora squeezes through the crowd to reach us, smiling charmingly at everyone as she goes. It is a wonder that no one challenges her for pushing through to the front, but they don't. Her bonnet has slid off her head sideways in the effort.

'Isn't it a grand day? Isn't it a blessing that the sun shines? Won't it be fun to watch? Do you think we'll ever see the like again?' She doesn't wait for any answers, nor does she take her eyes of the southward track along which the foreign dignitary will travel. 'We'll get ever such a good view, won't we, John? He'll be travelling in the boat to inspect the bridge, and we'll see him close up. Oh, they say his clothing is ever so exotic, John!' Her voice squeaks with excitement and Bristles O'Malley frowns and mumbles more revolutionary curses into his beard.

'Ohhhh! Here they come!'

But we didn't need Cora to point it out—the royal train has come into view, and it feels that the whole jetty rises as everyone get on their tiptoes. As the carriage approaches, the crowd begins to cheer, with the noise travelling along the track. 'Hurrah! Hurrah!' I join in, waving my cap and jumping almost as high as Cora beside me.

'He is not even looking!' my friend protests in outrage.

It is true—the Shah and his companions in their exotic fabric crowns do not pay attention to their cheering audience. I catch a glimpse of the Shah talking with Sir John Fowler, the designer, through an interpreter. The carriage slows and halts.

After what feels like a small eternity, the esteemed party disembarks and they make their way to the *Dolphin*, the best boat in the construction fleet. We onlookers swivel sideways to keep a good view. For a moment, the sun catches what looks like a golden teapot and Cora beside me gives a sharp intake of breath.

'How luxurious!' she breathes to no one in particular. The boat makes its way beneath the bridge, with the Shah sitting cross-legged beneath a crimson canopy. 'He has neither looked left or right. Can nothing amaze that man?' Cora asks, tugging at Bristles O'Malley's sleeve. He huffs and shakes her off.

Suddenly the Shah in the boat begins to point animatedly and is on his feet. A shiver of excitement passes through the crowd. A cage of workmen is travelling vertically up on the structure, but no, that is not what has caught his attention.

'What is he pointing at?' Cora's question is echoed by perhaps twenty others, standing cramped on the jetty with us.

I narrow my eyes and shade them against the bright light. 'Oh, goodness!'

'What?'

The fluffy russet tail and the graceful bounds along the girder are unmistakeable.

The Shah of Persia has taken a shine to my squirrel!

As is his way, Rusty disappears between the steel struts reflecting the sun.

It is only when the *Dolphin* returns to the jetty, that we can fully appreciate the Shah's appearance. 'How can he be needing such a thick coat on a day like this?' Bristles O'Malley mutters while chewing on his unlit pipe. He is right. The Shah is wearing a fur-lined, emerald-studded cloak and casts his eyes up and down the assembled audience through the gold-rimmed spectacles attached to his hat.

'Now that you say it, it's warm today!' I laugh and begin to wriggle out of my jacket.

And then two things happen at once. I feel Cora's hand shoot over to my chest pocket with purpose and extraordinary speed. The next thing I know, my jacket lies on the floor and Cora has jumped the barrier, dodging the Shah's panicked servants.

18

Who Dares Wins

She comes to a halt before His Imperial Majesty and Sir Benjamin Baker and holds my book up in front of them both. They look at each other.

'Cora! Give it back!' I shout.

But my friend ignores me. She curtsies and reaches into her skirts, retrieving a fountain pen and a small bottle of ink. I cannot believe what I am seeing!

'CORA!'

My call is copied by Cora's father Mr Ramage far behind me, who has only just realised now what his daughter has done.

The Shah and Sir Benjamin lean forward to speak to her and the crowd jostle. I can't see! Bristles, Doug, and Jim have been pushed sidewards, resulting in more men edging forward and knocking me off balance.

Cora smiles sweetly, curtsies to the Shah, and to every last one of the assembled gentlemen for good measure and is ushered back towards the barrier.

'I know. I was excited. I'm sorry,' I hear her say to the policeman who escorts her back. But in her hand, she holds my open autograph book.

I reach to snatch it back, but she pulls it just out of reach. 'In a minute, John, not yet. Wait until the signatures have dried.' Her smile is an arch of triumph.

I stare down—Sir Benjamin Fowler's signature is almost dry. But on the next page just across is a strange looking set of lines and curves. 'Is that really…?'

'You owe me, my friend. The signature of His Imperial Majesty the Shah. I figured that we might as well ask. Who dares wins, isn't that right?'

I am speechless.

***

I am not the only one. The library is closed by the time I get there, but I know that Mr Peebles lives in the flat above, and I ring the bell without a moment's hesitation.

The librarian appears at the side door looking a little dishevelled, having loosened his necktie, and there are crumbs in the corner of his mouth, but I pay these things no heed.

'Mr Peebles, you won't believe this. Look!'

I push the book under his nose. Then I think better of it, in case some of the crumbs in his beard fall on the page and spoil the signature. I hold it an arm's length from the librarian's face. 'Oh, Mr Peebles! Imagine my luck! The signature of the Shah of Persia, do you see it?'

His mouth falls open. I hold the book out, but it's as if the real Mr Peebles has gone into hibernation for a moment. The sky is still bright, as summer nights are in these parts of the north. And just like that, he is back. 'John, this is extraordinary. How remarkable! How wondrous! Compliments, my boy, and hats off to you.'

This, precisely, is why I love the librarian so much. I drink in the wonderful words he uses, soak them up like a tree soaks up rain.

He cannot take his eyes off my book. 'John, let me find out how much this book would sell for if you were to decide to part with it. If you continue like this, you'll be a breadwinner for your family without ever scaling a building site again. Hmmm, Carnegie, Stevenson, Fowler and the Shah. I'm going to give this some serious thought. Somewhere, I read about an auction where an autograph book like this one sold for an awful lot of money, though that may have been in America. Hmmm.'

'I should go home.' I state carefully, supressing a grin. 'I'm on the early shift again tomorrow and they tell me young men like me need our sleep!'

The librarian peeks over the rim of his glasses, a small smile creasing the corner of his eyes. 'True, they do. I'm just pondering how you can possibly improve on this. The signature of her Majesty the Queen, perhaps?' Now he laughs and I join him. 'I think that is, we'll both agree, impossible.'

'As I say, leave it with me. I will make it my business to find out what this could mean for you and your family.'

As I lie by the warmth of the range at night, images of the day float across my mind like the morning mists on the Forth, thinning and thickening and then dissolving like they were never there in the first place. The squirrel on the bridge, the Shah's excitement—and the moment of madness when I watched Cora scale the barrier for nothing but friendship.

I didn't even thank her properly. I must put that right tomorrow. I pat the autograph book under my pillow three times before I breathe in and out deeply and give myself over to dreaming. A house without the threat of bailiffs, a life without hunger, an education for my sister—I am not greedy enough to think I deserve it all, but I would love a little of it, a slice of the life which others take for granted. Before I snuff the candle out for good, I say a heartfelt prayer of thanks and vow to value Cora more. Is there anything I could do for her?

***

The next morning the train is still buzzing with yesterday's events. I've decided to leave my autograph book at home for a day or two. I'm not even sure why—perhaps I am more scared to lose it now. I will wait and see what Mr Peebles discovers. Perhaps the autographs will fetch enough to get us to a better house, instead of sharing with so many other families. Mother is back at work in the linen factory now that Dey is at home and can look after Helen, but he is ailing, and it worries me.

Word of Cora and her antics has spread, and one or two of the briggers on the train acknowledge me. One says: 'Your friend is a wild one, that's for sure.' I answer with a wry smile. *I know she is.*

To our surprise and consternation, we are ushered off the transport boat and marched along to the Queensferry pier where MacOrderly stands waiting to address us.

The men begin to mutter amongst themselves: *What's happening? Why aren't we allowed to start work?* Stragglers join from all sides until hundreds of us crowd onto the space. All of us keep our voices low to hear whatever the man in charge may have to say.

There are those who eagerly expect bad news, welcomers of woe like Bristles O'Malley, kind though he is. Then there are others who would never countenance such a thing and expect greatness around every corner. The truth is always somewhere in the middle.

'We're in trouble, so we are, I'm sure of it,' mumbles Bristles into his beard. There must be hundreds of men and boys here by now.

MacOrderly stands on a makeshift box to address us.

He begins. 'The Shah's visit was a great success yesterday, that much is clear. However, there is something that gives us great concern. As the Shah and Mr Baker's train neared Queensferry...' MacOrderly hesitates, and his voice becomes strained. 'As they neared here, an obstacle was discovered on the line. Someone, somewhere, had seen fit to roll rocks onto the tracks.'

A man from the back pipes up.

'Couldn't there have been a landslide? Who says it was foul play?'

'Well may you ask!' MacOrderly answers drily. 'There was a message daubed onto the sleepers. Do you want to know what it said? It said: *The bridge will fail. We will see to it.* No, this was intentional. Sabotage. We're looking into it.'

Further murmurs spread through the crowd of briggers as the reality sinks in. *A steam train in full flow would have derailed with such an obstacle. If the Shah's train had careered into the rocks at full speed... No, it's simply too awful to imagine.*

Whoever sabotaged the track was willing to risk murder.

19

Mind How You Go

The summer months pass by, alternating between heat and haar. The sun burns off the fog by lunchtime and the air is crisp with success. Every day I climb up and up while the bridge's cantilevers near each other more. Soon the central girder will be placed, joining the parts of the bridge together. This is the last summer of working on the site, MacOrderly says. We're on the final stretch.

I still feel a flutter of fear on the way up sometimes, but the terror diminishes quickly. I do look forward to each day with Bristles O'Malley, Doug and Jim. We work hard and we enjoy each other's company. They delight in Rusty coming onto the bridge, which he now does most days, and everyone throws him food, but I am the only one whose shoulder he sits on. He doesn't linger long—it's the wildness in him—but it is enough for me: a fleeting brush of the soft fur of his tail, then the red streak along the girders in the distance as he makes his way back to land. I love his tufty ears which move forward and backwards so eloquently that I can often predict what he is going to do next. The men call me the squirrel whisperer and guffaw heartily at me, and I laugh with them.

My birthday has crept up on me and I cannot believe that I have been a brigger for a year. Mother is already at work, and wee Helen sits on the kitchen floor and plays with my old marbles. I secretly still like them, round and cool in my hand, but if I'm a breadwinner, I should really move beyond things like that.

'Oh no, it's Friday 13th,' rumbles Dey from his bed from which he barely rises these days. 'Are you no working the day, John?' Dey is getting a bit forgetful too.

'No, Dey. My gang leader Mr O'Malley burnt himself yesterday and has to take a day off to let the burn settle. All of us will be back tomorrow, God-willing.'

'Right, son.'

'But it's my birthday, remember Dey? I am going to the library to see Mr Peebles.'

Dey shakes his head. 'Mind how you go, John. Friday 13th is unlucky, I'm sure of that.'

'I'm not worried, Dey. Really, I'm not.'

I cheerfully pull the door closed behind me and take a walk along the road towards the town centre, whistling as I go. Pittencrieff Park stretches out towards my left, the Tower Burn tinkles in the distance and the ruins of Dunfermline Abbey glisten in the morning dew. I lengthen my strides and all but run the last few steps uphill to the corner and the entrance to the library. I must have grown—Dey's jacket sits tighter on my shoulders.

'Good morning, John,' whispers Mr Peebles as I step into the lending room, but his eyes remain on his boy assistant who has dropped a big volume while re-shelving. The spine of the book is bent, and the back cover hangs off at an angle. Nevertheless, the assistant simply slides the book into its place on the shelf, stealing a sneaky glance back. He may have hoped that Mr Peebles did not witness his misdemeanour, but now he has another thing coming.

'Oh, for the love of words!' hisses the librarian, dropping his monocle and rising from his seat. 'That book is valuable! Bring it here and be quick about it.' He doesn't shout, but his disappointment is evident in his sloped shoulders and his deep breath. He gently receives the broken book from the hand of the culprit who doesn't even apologise. The boy assistant saunters back to his trolley to resume his duties. Mr Peebles' forehead wrinkles and he sighs, but he stops short of complaining further to his helper. With a wink, he turns to me. 'Perhaps there is truth to what they say about Friday 13th being unlucky,' he says, clearing his throat, adding in a whisper: 'He has dropped three books and knocked over a customer this week.' The librarian shakes his head. 'But you look cheerful, John!'

I flash him a smile. 'It's my birthday. And I am here which is my favourite place, but you already know that.'

The librarian returns my grin before leaning towards me. 'Anyway, as promised, I have looked into it all. And I have some news for you, John. Naturally, it is difficult for anyone to predict exactly what your autograph book would be sold for should you allow it to be auctioned, but others of a similar type have fetched up to three hundred pounds and my contact at Lyon and Turnbull's auction house in Edinburgh predicts that yours could be no different. We'd have to travel to Edinburgh, I suppose, but I can make enquiries. You have something special there, John.'

My mouth hangs open. *Three hundred pounds?* I do not pretend to know much about money, but one would have to work a long, long time as a brigger before making such an amount. *I have to tell Mother! And whatever will Cora say to that?*

Passing the pillars at the exit, I plan to head to the baker's shop for an iced birthday bun when I am stopped in my tracks by a commotion outside. An old man and two ladies talk in urgent voices.

'Not another one! That bridge! It never stops, does it? Never enough. Poor Aitkens.'

'17 years old, too. I mind John, right enough, pleasant lad. Killed outright. They say he fell with carelessness in the fixing of a stage upon which he was working.'

'Crying shame, so it is. And today of all days.'

They share silence, and a shiver.

I only realise now that I have been standing rooted to spot for the entire conversation, staring at them from no more than two yards away.

'Sorry, laddie. Did you know the boy too?'

I pass them with a mumbled *excuse me* and break into a jog. Ten paces later I am running, every muscle tensed to the point of pain. Another one dead, and a boy, just like me. On the thirteenth of September, my birthday, just like David Clark last year. Bile foams in my mouth. The sun retreats behind the clouds and an icy chill descends, which does nothing to dismiss my doom. I sprint all the way to the train station—and as I suspected, the midday train is just about to pull away.

I don't want to see Dey, or Mother, or even my old friends from school. How could they possibly understand? No.

I need to speak to Cora.

20

Give Yourself a Shake

When I make my way to the works across the water, there is an unmistakeable hush. MacOrderly's office lies abandoned. Some way off, important men in top hats and greatcoats observe a moment of silent reflection, or at least that is how it looks. I can see Cora's father's boat out on the Firth, but it is too far away to tell whether she is with him.

John Aitken's smiling face hovers in my memory, just ahead of me in training in the sheds. The bridge towers above me. I hate it. Hate it for what it's doing to the Clarks and the Aitkens and scores of other families. What it may yet do to Mother and Dey and Helen. And yet, somehow, I want to be here today. This is my life. This is my place. These are my people. I don't know why, but I walk, along the pier and up the first ladder, then the next and then the next after that.

I wait for the lumps to rise in my throat, the shaking of the hands, but none of it comes. I am numb. It could have been me. Perhaps it *should* have been me.

I take a walkway to the left and another two ladders lead me upright, past rivet gangs hollering into the wind. Overhead, gulls shriek and clouds obscure the very highest part of the structure. I realise—that is where I am going—the lonely place above it all. If men are working to the right, I turn left; when men are working to the left, I turn right. Only upwards, until I am among the birds. The steely limbs of the bridge lift me to a place where I can see clearly. Beneath me it creaks and shifts and grinds in the constant rhythm of sea and sky, wind and waves, and I allow my body to move with it.

I shade my eyes to observe the wide expanse of the Firth of Forth, with the Queensferry and the Inchgarvie cantilevers straining to reach each other. Soon, the three cantilevers will form one structure. In the distance to the north, south and west, hills rise in a circle like a crown around us. I stand freely, raise my arms, and let the wind shake the grief and the worry from my clothes, breathing deeply until a strange calm comes over me. Lower down, the hammering and drilling and welding goes on and on and on, but up here I am at peace.

For a long time, Cora's father's boat bobs like a stranger's toy in the distance. Only when it pulls towards the Inchgarvie pier, I snap out of my reflection. *Wake up, John. Aitken may not have survived, but you have. God saw fit to spare you, and you must do the best you can with what you have been gifted. Make a better life for your mother and sister. Be a good worker and a good friend and make something of your life. It is your birthday! Give yourself a shake, John Nicol!*

I hop, slide and leap down past all the industry on the bridge until, slightly out of breath, I stand by the pier waiting for Cora's boat to come in. I am not disappointed: she is there, leaning over the gunwale.

'John! I thought you were off today.'

'I am.'

She looks at me, not quite comprehending. But then her expression darkens. 'You heard about the Aitken boy.'

I nod, catching the rope from her father and tying it up as he showed me. 'People were talking about it outside the library, and I came straight here.'

She skips to shore and inclines her head sideways. 'Papa has another run to do, taking equipment across. I have time.' She walks towards the water's edge and ruffles her dress aside until she can sit comfortably, dangling her legs over the water and staring into the depths.

'Did you see him fall?' I ask tentatively.

She hesitates. 'No. But there was a great uproar at once, and I saw that.'

I chew my lip. 'Cora, today is my birthday, and the thirteenth. The boy from my school fell exactly a year ago. And today, Aitken. And my father died before I was even born. Cora, do you think it's possible that…'

She suddenly grins. 'You think you're cursed? John, I may be a girl but I'm not an idiot. That sort of superstition is nonsense. There are no curses, only blessings and trials—and we must choose how to respond to them.'

She tosses her hair sideways for emphasis and stares over the water while I try to make sense of it. Already I am feeling lighter. Just talking to a friendly soul lifts the worry wisp by wisp, until it is dispelled altogether. I attempt a smile back and to my surprise, I succeed.

'Do you want to hear something cheerier?'

She listens attentively as I repeat my conversation with the librarian to her.

'That is wonderful, John! What did I tell you? Blessings! Not only trials. This will be the making of you, mark my words. Oh no wait, I didn't know it was your birthday! I obviously haven't got a present for you, but there is something I have been working on. It might not come off, but here goes.' She tips two fingers together and loops them into her mouth to whistle a low, powerful burst. She repeats it. I wrinkle my forehead. Every few seconds or so, she smiles and looks along the structure, waiting for something to emerge. And then it does, a rust-coloured streak of dark reddish-brown, a swish of the fluffy tail.

'Rusty!' I call. I haven't seen the squirrel for days.

'I think I may have trained him a little. A whistle doesn't always work, but…' Cora trails off throwing the animal a small handful of breadcrumbs. 'He's quite tame now, isn't he?'

And don't I know it! I freeze still as I allow Rusty to clamber up my arms, sit on my shoulder and jump onto my cap where he waits expectantly.

'Happy birthday, John. Mark my words: you are blessed, truly. You only need to open your eyes and see it.'

Rusty has scrambled down my back and crouches in front of me now, looking and listening.

'Gosh, I have to go!' Cora smiles and rushes towards the pier where her father's boat is pulling in again. *Soon it will be time for the train to Dunfermline. And I'd better be on it if I value my life. Mother would not hear of my staying out over teatime today.* I catch the next transport boat and run towards the platform as if I'd just come off shift. In the distance, twigs quiver under the featherlight weight of Rusty as he returns to his home in North Queensferry's trees.

All the chat on the train concerns the Aitken boy. How easily it can happen. How fast, how deadly. A year ago, this would have made me shake in my Dey's boots, but today I climbed to the top of the bridge and defeated that fear for good. I feel stronger somehow.

As I lie in the darkness by the kitchen range at night, I count my blessings just like Dey taught me. However, it isn't long before I wonder, like I do most nights:

What has become of the Murdochs?

21

We'll Cross That Bridge When We Come To It

By the end of the month, the sides of the Queensferry and Inchgarvie cantilevers are so close that someone suggests securing crane lifts to both sides and balancing a ladder between the crane jacks. I am almost oblivious to my fear of high places now, but my heart clenches all over again as I see a volunteer crawling along, two hundred feet above the water. He is the first to cross, and once a makeshift gangway is in place, the Hawes Inn rings with singing that night.

Two weeks later, the gangway is replaced by a more secure crossing, and by the sixth of November, the central girder is ready to be connected. Instead of three cantilever towers stretching towards each other, our bridge will soon be one structure, ready to carry train tracks across the Forth. The end is in sight.

More and more people stop by the site on sunny afternoons, and the briggers enjoy the diversions which come with the visitors. On one occasion, a showman and his troupe set up a puppet booth by the shore. Its sign reads *Professor Moffat's Royal Entertainment Show*. All of us are mightily impressed by the 'royal' part of it, but Bristles shakes his head and walks away to admire their dancing bear.

The bridge is taking its final shape while the Black Squad carry on laying the rail track. Meanwhile, we continue with the riveting work, but there is no denying it: the site is emptying of workers. Less hollering and hammering, and more than a few leaving celebrations as the workshops reduce their workforce and man after man packs his knapsack to return wherever he came from. I am glad of it. By now, the winter winds are on their way. Whenever I pass the North Queensferry hillside, I scan for glimpses of my squirrel, digging and hiding food. Rusty comes to the bridge less than he used to. I am just thankful that the train all that time ago did him no lasting damage.

After all work is called off due to iced ladders and biting frost, we're released for Christmas.

'John, son.' Mother's voice is jolly, but there is something in it I cannot put my finger on. Something foreboding. She places two bowls on the table. Dey is in bed as usual now, and Mother looks tired. I can hear Helen from the street below where she is drawing patterns into the snow with sticks. 'John, I have been told I need to find a new position.'

I look up from my steaming plate of stovies.

'Don't worry, I am looking. But money may be scarce for a while, and with Dey's doctor bills…'

I know what she is saying. My brigger work will come to an end, too. And soon.

She smiles a tight smile. 'There are plenty of factories and I'm a good worker, I know that. But I think we will have to move to a smaller house, John, and I thought that since you're a breadwinner, you deserve to know the truth.' With that, she tucks a strand of her hair back into her bonnet and resumes her meal.

Mother is a marvel. Others may have cried or complained. She ladles out an extra portion of stovies and hugs me instead. 'There's plenty folk who have it worse than us, make no mistake. We'll manage.'

I return her smile with all my heart. There is misery enough. I will have work at the site until March. After that, only the Almighty knows, and that will just have to be all right with me. I take a deep breath and pat the autograph book in my pocket. *You never know.*

On the 21st of January, Cora and I shelter in the porch of the Hawes Inn. A gale has been blowing all night and hail pummels the roof slates from time to time. All the remaining briggers have been ordered off the bridge, due to the weather I presume. My stomach rumbles as it often does when I have nothing to do. Cora has a science book in her hand. I wish I had thought to bring my library novel—I have discovered Mr Dickens. Suddenly, there is a commotion inside the inn and ten or twenty people spill out onto the road and the pier. And finally I hear it too—a low rumble.

What is going on? All up the Queensferry hill, doors and windows open with excited shouts which are soon eclipsed by an approaching noise.

'Ouch, Cora!'

'Sorry!' she lets go of my jacket sleeve into which she has been clawing without realising. She jumps a couple of yards away from me and reddens, shaking her hand as if she had touched something revolting. Soon neither of us can think about that anymore. The sound is deafening, passing overhead.

Trains! For the very first time! Trains, two abreast, passing together along the tracks, across the viaduct above us and then onto the bridge itself. The first steel tower is alive with rhythm, sending vibrations through us all. Two sharp whistles pierce the wind as the trains edge forward slowly, leaving steam plumes in their wake.

'Did you know this was going to happen, John?' I can barely hear Cora, because a new bout of hail is drumming onto the rooftiles and bouncing off the road. I shake my head in answer. We observe, all of us, as the two trains make their way along the Queensferry cantilever. A particularly powerful gust of wind worries us for a moment and Cora is by my side again. 'John, I bet they are doing this today *because* of the weather, not in spite of it. Remember the Tay Bridge disaster? If they test our bridge and it can carry two full length trains in conditions like these—well, then it's safe for any conditions, do you see?'

'I suppose.'

I am distracted. The trains have stopped—both are still on the bridge, but closer to North Queensferry now.

'What could the matter be?'

All around us, men and women ask themselves the same question. Finally, a young man with an impressive moustache arrives and speaks up. 'Ladies and gentlemen, nothing to see anymore. There seems to be a temporary problem with the track at North Queensferry. The rail workers are seeing to it, but they suspect the weather caused some of the tracks to come loose. The experiment has been a success!' The adults nearby clap and cheer.

'Could it be sabotage?' I mumble under my breath. 'The loose tracks?'

'Sabotage?' Cora asks, but as she says the word aloud, her face darkens with understanding. 'Are you still on about that family? I thought they had moved on.'

I hold her gaze. I wish I could believe it. Behind us, the moustached man ushers the onlookers back into the inn and Cora's face lights up with recognition. 'Gosh, John. That's Ernest Moir—the engineer in charge of this whole cantilever! Get him to sign your book! He is married to the lady I showed you, remember?'

The young man with the moustache may not be all that famous yet, but who am I to say that he will not make his name in years to come? I fumble in my breast pocket as I stumble towards him. And there she is: Margaret, his young wife. 'Sir, could I trouble you for a moment,' I begin with a croaky voice. 'Congratulations on your success, Mr Moir. May I ask a favour? I would be ever so grateful if could sign my book. I have made it my business to collect a few signatures to pass the time and I'd be very glad to have yours.'

I attempt to shield the pages of the book against the wind and rain. The last thing I want is for one of my famous signatories' names to be smudged or illegible. Thankfully, he has a pencil in his pocket, and a few seconds is all it takes. His smiling young wife stands beside him, staring at the trains fading into the distance above the water. She notices Cora standing by my side. 'Don't stare at *me*, girl,' she laughs. 'Look up. One of those trains up there is driven by a woman. Lady Tweedsmuir. Can you believe it?'

Cora gasps and blurts out: 'Is it really? I'll have you know, I long to be an engineer when I am older. I build things at home all the time. I take things apart and put them back together too—clocks and so forth.' My friend claps a hand over her face, as if her head had just caught up with her heart's outpouring.

The lady shrugs. 'Ah, I'm afraid the men have the engineering in hand, my girl, though we can advise them...' Mrs Moir smiles patiently.

'That's not enough for me,' mumbles Cora as the lady leaves. Lady Moir turns and locks eyes with my friend for some time. Then, slowly, the lady nods. 'What's your name?'

'Cora Ramage, begging your pardon Ma'am. I spoke out of turn.'

Still, the lady looks at her with an intensity I find a little unsettling. Her voice is quiet, thoughtful. 'I don't think you did speak out of turn. Perhaps one day. Perhaps, if enough of us want it, we can make it happen. Imagine, Miss Ramage: *The Women's Engineering Society* or some such.'

'I hope so, and soon,' Cora says with feeling, before buttoning and unbuttoning the collar of her coat in embarrassment.

'Margaret, are you coming?' The man with the moustache is holding the door. His wife follows, nodding farewell to us both.

'How do you do it?' I ask.

'What?'

Seize the moment like that, spotting persons of note! If this book fetches good money, I'll need to give you a share!

'No need,' Cora says blithely. 'I'll be rich, remember? I'll be an engineer!'

22

The Golden Rivet

MacOrderly's assistant Elizabeth from the offices hands out leaflets as we all arrive at the Queensferry pier. Bristles O'Malley takes the paper but immediately crumples it into his trouser pocket. Doug is not paying attention— he is trying to catch Elizabeth's eye and smiles broadly when she gives him a shy wave. I swear I saw him skip just there.

I sidle up to the Irishman. 'Mr O'Malley, what did that piece of paper say?'

My gangmaster ignores me, which is unlike him. I try again. 'Mr O'Malley, it might be important, if the people at the office are taking the trouble to hand it out to all the rivet gangs.'

Jim Smith and Doug are veering off to pick up the sacks of rivets for today's shift and O'Malley beckons me over.

'Here, don't be telling anyone, now. But I can't read it.' Bristles O'Malley mumbles the words into his beard—

and they land like a blow to the stomach. *He can't read!* I feel such a sadness for this poor man. I don't know where I'd be if I didn't read every day. My latest favourite is Mr Dickens again, *Oliver Twist*. I hold out my hand and Bristles O' Malley places the crumpled leaflet into it. I begin to read aloud:

All remaining rivet gangs will be entered into a draw to assist His Majesty the Prince of Wales as he secures the symbolic Golden Rivet on the day of the grand opening. Winning gang announced tomorrow.

I read it twice over, louder the second time so that Doug and Jim can hear too. 'What an honour that would be!' I add when I'm finished. Bristles O'Malley shakes his head. 'It's a fierce nonsense altogether, so it is.'

Despite his grumbling and mumbling about our Crown Prince, the rest of us sing *God save the Queen* loudly throughout our shift, half to tease our gangmaster who can't help laughing in the end, and half to boost our chances of winning.

'Somebody has to get the honour,' I tell Doug as I bid him farewell at the end of our shift. 'It might as well be us.'

Despite the excellent progress the project has made, there are a thousand things still to do in the run up to the grand day. I am lucky that our rivet gang still has work to do. The huts in Queensferry are emptying. Once or twice, I fancy I see Murdoch-shaped figures in the distance, staring down from Dalmeny or strolling past the pier. I never get a good enough look, and in any case my imagination alone is enough to paralyse me with fear.

'Will you leave off, John!' chides Cora. 'They have gone. Briggers move on all the time; they go where the work is. Relax.'

If only it were that easy.

'Forget about them, John. Besides, have you seen the newspaper reports about the opening? There are all manner of famous persons attending,' declares Cora in a loud voice when I lower myself from the final ladder at the end of my shift. 'I heard two gentlemen talk about it on my father's boat, and one left his paper behind! So, it's true: His Royal Highness the Prince of Wales is to open the bridge and drive in the final rivet, but they are using a golden one so it's soft enough not to require heating. And the Prime Minister is coming too, and lots of famous guests of honour.'

I allow myself to dream for a moment, and then sigh. 'I am not likely to get near them, Cora.'

But the next morning, MacOrderly himself stands on the pier, an uncharacteristic smile playing about his lips. We all file past him, but he puts his arm out to stop Mr O'Malley on his way onto the bridge.

'Mr O'Malley. A word, please.'

I catch Doug's eye, he elbows Jim, and we all stop

to watch, just long enough to see MacOrderly say 'Congratulations, it's a great honour,' and shaking Bristles O'Malley's rough hand.

'It's us, isn't it?' yells Doug, his floppy fringe bouncing with glee. 'Isn't it, Mr O'Malley?'

Our gang leader catches up with us and shakes his head in mock disdain, although it's hard not to share Doug's expectant joy. *Did Elizabeth from the offices pull some strings on our behalf?*

'It's us,' Bristles O'Malley finally grunts, and Doug, Jim and I whoop and perform an impromptu jig right there on the pier, with the rest of the workers looking on and clapping.

23

Keep Your Friends Close and Your Enemies Closer

It's another dreich kind of day with splatters of showers sweeping across the Firth, but by mid-morning, the crowds begin to arrive. Soon, both Queensferry and Inverkeithing are overrun with spectators—they pour out of special trains or arrive on foot, or by bicycle. By noon I wonder where more spectators could possibly go.

I have made it my business to be early, despite the day off that has been declared for all briggers. Cora's father allows me a ride to Queensferry where most of the festivities will take place. Mother and Helen plan to walk from Dunfermline to the northern shore of the Forth to catch a glimpse of the royal party, though I am not sure how they will. Dey could barely speak with emotion when I told him that I am going to help the future king drive in the last rivet.

In the distance, I can see police and volunteers ushering spectators this way and that, and I am glad to have arrived so early to secure my spot on the pier. Up on the hillside of the Hawes Brae, locals and visitors stake out their viewing spots and doors and windows open wide despite the weather. I look at the crowd closely. But, no, all my fears of sabotage are misplaced. I breathe a little deeper. The bridge is finished. Prince Edward will soon be here. In the moody waves of the Firth, the huge HMS Devastation comes into view, surrounded by smaller boats. I glance up to the well-dressed ladies and gentlemen who have been given permission to watch the proceedings from the footpath on the actual viaduct above us, although I do not envy the ladies with their bulky skirts and rustling hats. Some look down with trepidation. Once, I might have been like them. It is a long way down. Soon the Prince himself will be among them, getting ready for the driving in of our symbolic last rivet. I hope our future king has a good head for heights.

Beside me, Cora holds on to my shoulder to push herself higher to see. 'I think the train is coming,' she exclaims. The crowd turns as one.

Except me. I feel the blood drain from my face. Behind the Hawes Inn and behind the turned backs of the entire town, I see them. I am certain—the way they move, the way the foggy daylight contrasts with their red hair. They are heading for the viaduct. If I am not much mistaken, one of the young Murdochs has got a coiled rope slung over his shoulder. 'Cora!' I rasp, but my voice is wiped out by three large cheers, each louder than the other. All around me, men and women wave their handkerchiefs as the Prince's train rattles and steams onto the bridge and out of view.

'Is that it? Asks an old woman along from us.

'Not at all,' Cora explains. 'Watch out for the *Dolphin*, the ship which will carry the Prince around Inchgarvie island. See, that's it coming into view, I think. I wish it wasn't so misty! But yes, that must be it—my father said it had been decked with crimson and gold to celebrate the royal visit.'

I tug at her sleeve. 'Cora, the Murdochs are here! They are walking onto the bridge. What is the Prince doing next?'

Her mouth becomes a tense line. 'Are you sure?'

'Deadly certain.'

Just at that moment, Bristles O'Malley, Doug and Jim appear and give me the nod. 'It's time John, so it is.' The Irishman is wearing a starched white shirt and a lopsided bow tie, and I swear the bow tie is stitched from his old cap. I don't mention it. Doug looks dapper with a proper jacket and his moustache all trimmed in honour of the occasion.

'How far is it?' I ask.

'Beyond the arches there, so it is.' O'Malley leads the way. Cora gives me a tense smile and I follow my gang up to the ladder that leads up to the viaduct pavement, directly beside the train tracks. However hard I stretch, I cannot see the Murdochs again and now I am wondering if I imagined it after all.

Most of the spectators are sheltering from the rain in the granite arches at the end of the viaduct, with only a few pressing on to the platform where the rivet ceremony is going to take place. Who are all these dignitaries allowed up there? And wouldn't the Murdochs look out of place? No, it is so crowded—and workers and messengers are running this way and that—no one would ask questions, of that I'm sure.

There! Oh no, I see them. Like fish swimming against the stream, the three figures make their way onto the bridge ahead of us. Goodness knows how they got past the barriers. Behind us, the sounds of Queensferry fade into the distance: musicians and singers in the streets, posh ladies spending their money on games and entertainment while their menfolk inspect the workshops and buy drinks among stall holders selling souvenirs of the day. I elbow my way forward, trying to keep the hunched figures in my sight. *Keep your friends close and your enemies closer.* This time I won't let them leave me behind.

'Slow down, John,' puffs O'Malley. 'The Prince won't be at the platform for a while yet, so he won't.'

I ignore him.

Ducking and running along the walkway, I spot them again. Most ladders have been removed but the Murdochs know the bridge like the back of their hand. Soon they do what I expected them to do: they disappear up one of the struts, two bays before the platform where the ceremony will take place. *This is bad. This is really bad.*

'Bristles, Doug, Jim. Trust me please. I'll be back.' I don't have time for more explanations.

I swing myself up on the lattice girder directly behind the strut the Murdochs are climbing. They won't be able to see me, but I have lost sight of them too. The waters of the Forth, and the ceremonial platform beside the train track, sink beneath me as I clamber up. It is incredible to me, but no one shouts my name or calls me down. If they do, I don't hear them.

Aside from my own rivet team, no one is looking up at all. Will Cora have the presence of mind to attract the attention of a police officer? I can't be certain, and I cannot afford to take my eyes off the strut beside me. Within seconds, they emerge from the hollow structure through a side hatch. One of the twins goes first, swinging himself into a narrow gap between girders. He secures the rope and allows his father and brother to make their way across, supported. I am gaining on them. Higher and higher they climb, and I follow.

Fragmented melodies of the national anthem float

across the water in wisps and I nearly lose my footing when a deafening crack rents the air.

I am not the only one to be surprised—there is a hearty curse from above and Canny Murdoch regains his footing on the steel girder. Another loud crack. Ah, HMS Devastation, the protection ship, is firing a 21-gun salute. No one is looking up. How can the crowds not see the three men, securing themselves with rope high on the structure? I only breathe deeper once I am diagonally beneath the Murdochs. Close enough to see the expression on their rage-twisted faces. Close enough to see the wind whipping their jackets. Close enough— *oh good Lord, help us all*— to see a gun barrel glint in the grey light of the day.

24

Courage is the Measure of a Man

I consider what to do. I could shout for help, alert the policemen below. But the wind is strong, the rain is already battering the steel and I may not be heard. Worse, the rascals in their hiding place may hear me all right, and then both I and the Crown Prince will be done for. I look around for a weapon. Anything. Anything at all. The Queensferry hillside is an anthill of writhing colour and noise in the distance. Wherever Cora is, I am on my own now.

For the first time in months, some of my old fear returns when I look down. But only for a moment. These lattices and steel tubes are my home now. I am not the boy I was, cowed by the bridge, intimidated by men. Resolutely, I peer over the side of my strut to see.

The villains have tied themselves to their hideout with rope, not far from Queensferry's central girder, Rusty's usual route onto the bridge. A particularly strong gust reminds me that, unlike my enemies, I do not have a security rope. I can only hope that the Prince's train will be delayed so I can work out what to do to stop them.

But time is not on my side and hope is not enough. The sharp whistle of a train carries on the wind. Gradually, the sound of the approaching royal train gets louder, and the metal structure vibrates with the untold power of steam. I hadn't thought about it, but as the train pulls nearer and stops, both the Murdochs and I are all enfolded in the steam cloud, and I praise my guardian angel that the March day is so cold. As soon as the air clears, I can see that the train has stopped. Is it Mr Arrol who is opening the door of the carriage and leading the royal party? Long greatcoats and black-as-night top hats, with hands holding on for fear they will get blown away. I scold myself for being distracted. What of the Murdochs? Seagulls circle above, momentarily unsettling me.

They have ducked out of sight, behind the metal strut. But then, like a silver witch's finger, the gun barrel slides out from the gap where they are hiding. Desperately, I cast about for a stone, a hammer—anything at all to prevent what will surely happen next. Here are a couple of disused rivets, the larger size! I weigh one in my hand. I would know how to heat it, but throw it and hit the gun's barrel? I do not know if I can. And what if it misses and falls, and hits the dignitaries beneath me?

There is no time to doubt though. I hurl my first metal

missile into the air. The storm carries it sideways into the metal some ten feet short of my target. It clangs loudly before hurtling towards the waves and the Murdochs' gun barrel withdraws in a flash. A quick glance down confirms that the royal party are none the wiser at all.

The rain begins to pour again, and the gale rises. The bridge resists, creaking and groaning with the effort. What are the Murdochs going to do now? I feel my throat tighten as the gun barrel emerges again in answer.

Who would kill a king-to-be?

Someone bent on revenge who has nothing, absolutely nothing, to lose. Someone who doesn't care if they live or die. They are determined to harm our royal guest, and I can't prevent them from here. I must get closer.

I swing myself further up on my rain-battered, slippery lattice girder and begin to slide myself along. There is no ladder or walkway to take me towards their hiding place, and I have no time to dither. I must distract them—it's my only hope of saving the Prince's life. I glance down. Below on the platform, some sort of key glints in the Prince's hand. Arrol is kneeling now, helping him, I think, with O'Malley and our boys standing by. The gun barrel is adjusted, pointing directly at the Prince. I have seconds, if that! And it is this desperation which lends me courage.

'Cain Murdoch! Canny! George! I can see what you're doing! You won't succeed!'

My voice and its volume startle me, but it is nothing to the surprise and confusion I have caused among the assassins in their hideout; I can hear it. *I have done it!*

Only then do I think it through: I am clinging to a slippery girder, high above the train track and higher above the water. I am picking a fight with three men and a gun.

I have just enough time to process what I am now seeing: the silver gun barrel swings round to point at me and me alone. The next moment is all sparks and noise and fire and pain, and my hands have let go of the girder, with my body sliding down diagonally. I clutch the side of my head, but the bullet all but missed me. There is no time to send a grateful prayer to Heaven—another shot ricochets off the steel and I have no choice but to allow myself to accelerate towards a join in the metal below. I land sure footedly enough, with a heavy thud, and I'm already running for cover. Crouching behind a metal strut, I breathe hard.

And then it comes to me. The solution.

25

What's For You Won't Go By You

It's a madcap idea, but it may just work. I insert my fingers into my mouth like I saw Cora do and summon all my concentration. A sharp note rises, and a hundred instant prayers accompany it. Will he come? It is the longest of long shots.

But then I see it, though I am too far away to hear the soft scratch of claws on the surface of the steel. A streak of red bounds along the northern cantilever and then onto the central girder towards the Murdochs. Just as I prayed, my wonderful squirrel leaps right between their feet. Cain Murdoch's harsh voice cries out in alarm, and his sons shout something in reply, though I cannot make out the words. For this brief moment, they are distracted by my squirrel, and that is all that matters. *Now, John!*

My father died due to a faulty rivet. Perhaps I can protect us all with this one. With all my strength, I throw my second rivet missile, close enough to find true aim this time. My rivet clangs hard against the villain's weapon and he drops it. The gun turns and tumbles through the air, crashing first onto the pavement beside the royal party and then into the black water below. Cain Murdoch himself flails, and before he can regain his balance, he topples off his perch, swinging helplessly from the rope which his two sons are straining to hold.

I throw myself onto my back in relief. I only notice now: the Forth below us has filled boats, with policemen and constables, pointing upwards, Cora in their midst. Between us on the platform, the Prince and his companions have hurriedly boarded their carriage again, this time to travel the short distance to the place where the Prince is going to make his speech, if he is not too shaken.

I am dog-tired. Only now do I notice a thin trickle of blood where the bullet grazed my skin. It drips onto my jacket, but I cannot afford to worry about that now.

I skid down the ladders rather than climb—but thankfully I can still trust my hands to hold me. Before long, I stand before a semicircle of constables. Some of the bridge's own security men are already high up, retrieving the Murdoch brothers from their hideout and Mr Murdoch from the air. Handcuffed and surrounded, they are led off to the police cart which runs alongside the tracks.

My mind is reeling as I sink onto platform where the Prince stood earlier and where the head of a golden rivet glimmers against the steel. Many people are speaking at once, but one high, determined voice stands out above all others: 'I am telling, you, sir, John Nicol is the hero— he spotted the Murdochs and prevented the disaster. He wasn't part of the plot; he foiled it!' Cora is positively shouting.

My rivet gang, my friend, the workers and the security men shuffle aside, and I detect the Glasgow accent of Mr Arrol himself. 'Where is this Nicol boy?'

Cora normally has difficulty stopping talking in mid-flow, but she does so now.

The police constable points to me. Mr Arrol approaches me and reaches into his own pocket, producing a clean pressed handkerchief. I gratefully press it against the graze on my head.

'I remember you, John Nicol.'

I struggle to reply, letting out a mixture of a nod and a cough. I barely have the energy to do that. Mr Arrol continues: 'His Royal Highness the Prince has opened the bridge now, although...' There is a twinkle in his eye, '... let's say it was a short speech! Now your friend Miss Ramage here has explained that you are a bit of a collector. John Nicol, I would like you to come to the banquet in the drawing loft this afternoon. It'll be a terribly posh affair, I'm afraid, but I am certain that I can make a few useful introductions and fill a few more pages in your autograph book.'

Cora beams from ear to ear.

The rest of the afternoon passes in a haze. I try to remember my manners, but what is the appropriate thing to say when the heir to the throne thanks you for your loyal service? How does one answer a French engineer who has just built a dizzying tower in the centre of Paris? Where does one look when shaking the hand of the Prime Minister, or Lord Kitchener, or the Emperor of Brazil, or King Leopold of Belgium, or the King of Saxony? Five hundred and fifty seats are set inside the drawing loft—I hardly recognise the space.

The toasts barely penetrate my brain, though I revive a little at the sight of the Waterloo Hotel's legendary lark pies. Over coffee for the grownups, the toasts continue, announcing knighthoods for the designers and engineers. Thankfully, little is mentioned of the briggers, and nothing at all of what transpired earlier on the bridge. I am grateful. Cora beside me talks animatedly with Lady Moir—there are few ladies present, but they hold their own in conversation. What on earth could they be talking about for all this time? I am worn out, clutching the pen Mr Arrol lent me, and my autograph book, now bulging with signatures.

The winds pick up even further in the afternoon so that, surrounded by his royal stalwarts, the Prince finally takes his leave. Construction dust swirls above the ground, the Hawes Inn opens its double doors to many a patron, but it is time: I gratefully take a ride across the water to Fife with Mr Ramage and Cora. Thankfully, trains to Dunfermline run all day today.

Mother looks dog-tired too, cradling Helen on her lap by the fireside while Dey leafs through the evening paper. 'Well, that was a day and half!' she says, with a sigh.

My stiff limbs feel as if they are forged from steel, riveted badly to my body.

I want to tell her everything that has happened, a blow-by-blow account of the enormity of what has passed. But I simply can't. Instead, I retreat to my corner and, run my aching fingers over the pages of my autograph book, appreciating my sacred treasure. The names in the book give me hope. *What's for you won't go by you*, Dey used to say. I don't know about tomorrow, but today I am spared. And that is enough.

The very next day, an express letter from London arrives, bearing the royal seal—and then I do have some explaining to do. Our neighbour Bessie tells half the street, who crowd in all the way up the common stairway to be first to hear of it.

It is a letter thanking me for my presence of mind and for my courage at a time of great peril. It is the letter of a mother thanking me for protecting the life of her son. It is signed *Victoria Regina*.

Queen Victoria herself.

26

God Helps Those Who Help Themselves

Mr Peebles and I disembark at Haymarket station where Mr Ramage and Cora are already waiting for us.

'Goodness, John,' laughs Cora. 'I've never seen you dressed like this!'

'It's all borrowed,' I begin, but Mr Peebles shushes me.

'Hush, John! You don't want to give away your trade secrets now, do you? The important thing is that we are all here. Now, I am not in the habit of attending auctions, therefore I must say I am a little excited. But nervous too. I am informed that one may inadvertently purchase something by accident, simply by scratching one's nose!' The librarian strokes his beard, and I am certain he must have run a comb through it earlier. I have never seen it so trimmed and tidy before.

Cora looks like a proper lady, carrying a thick book under her arm. '*Modern Steam Practice and Engineering*?' I read aloud. 'Sounds a bit heavy to me.'

'Mrs Margaret Moir let me borrow it,' my friend explains.

'She insists on taking it everywhere!' rumbles her father, but she ignores him, whispering: 'You see, John, one never knows when one might have a moment to read!'

I laugh, but Mr Peebles has looked at his pocket watch and drains of colour. 'Make haste! Everyone, onwards. That way.'

We follow him towards Princes Street. I have only been here once before as a small boy, but Edinburgh Castle looks just as magnificent as it did then, towering over the town like a faithful watchman. Dey told me about the extinct volcano beneath it. Today, the rockface glints sharply in the sun. We turn left towards Charlotte Square and George Street.

The librarian puffs. He is struggling to keep up with Mr Ramage and us children, all of us much more used to exercise that he can possibly be. 'It is number 51 George Street; I am told that we can't miss it.'

He is right. Many gentlemen and a few ladies are already lining the pavement before the impressive building of Lyon and Turnbull, the famous auction house. My throat goes dry. 'Mr Peebles, are you certain that children are allowed in?'

'I wrote last week to explain the situation,' he answers absent-mindedly, fingering his pockets for his letter to show the doorman. 'This should do it.'

A curt nod and a wave and we are through the door, into a bright hallway. Polite chatter filters towards us from the auction room.

We are seated close together, and it gives me great comfort to see that the librarian and Mr Ramage look almost as uncomfortable as I feel. Mr Peebles sighs as he sinks into his chair. But a second later, he sits upright again, wagging his finger: 'As I said, remember, stay completely still. Don't scratch, twitch, or itch. Don't even shuffle in your seat once it begins, or you might buy something you cannot afford!'

At this, Mr Ramage's hands fly to his face. In addition, he crosses and uncrosses his legs this way and that, ruffling his hair, rubbing his nose, stroking his beard. It's as if he is trying to get it all out of the way ahead of the event.

Rustling of skirts and crinkling of hats, squeaking of chairs and tapping of pipes—this is not my world. But I have my friends with me and what else could I possibly need?

The large door to the hall creaks shut. How many people are here? Two hundred? Perhaps more?

The first item up for auction is a painting. Pleasant though it is, I cannot see why anyone would part with so much money for it, but what do I know about art? I concentrate on breathing deeply. *Don't move. Don't move.* My nose has begun to itch.

A writing desk which belonged to a cousin of the Queen. *I want to scratch so badly!* An old sword, a casket or some such. *It's unbearable!* And now the tell-tale tickle of a sneeze is building in my nostrils! I bite my lip hard until it stops. The air is getting stale, and I just reflect that I'd rather be high on the bridge than trapped between the well-to-do here when Mr Peebles' foot gently taps against mine. 'Yours is next.' His lips barely move, though he jumps when my loud sneeze tears through the room the very next moment. Disapproving glares follow.

'Ladies and gentlemen,' begins the auctioneer. 'Our next item is a remarkable little book. Though a little scuffed around the edges…'

I shuffle uncomfortably in my seat.

'…it contains an astonishing number of verified signatures to rival any of the autograph sales in America. A mixture of letters and direct inscriptions, this book contains the handwritten names of Prime Minister Gladstone, author Robert Louis Stevenson—formerly of this city, French engineer Gustave Eiffel, Forth Bridge designers Fowler and Baker among many, many others. Unbelievably, there is also the signature of the Shah of Persia, Sir Andrew Carnegie (an audible gasp rises from the audience), His Royal Highness the Prince of Wales…' The auctioneer raises his voice over the crowd to deliver the final trump card, 'And our reigning monarch, Her Majesty, Queen Victoria.' Shouts of surprise are accompanied by bursts of applause. I take the opportunity to finally rub my itching nose.

'I invite an opening bid of...'

The bidding commences. I barely have time to glance this way and that, taking care to keep my head still, as gentlemen tip their hats and wave their papers. All of this appears to mean something to the auctioneer. The figure keeps rising, through the hundreds and up and up and up, but it's all Greek to me. Mr Peebles' knuckles shine white as he clutches the day's programme. I jump when the auctioneer's hammer slams down.

'Sold.'

I feel a sharp nudge into my side. 'John! John, stop staring! You look frightful!'

It's Cora. 'It has been sold to that gentleman over there. Look, he is walking forward now.'

It is true. The well-to-do man reaches inside his wallet and writes a note, handing it over to the assistant. The auctioneer keeps rattling on with his speech at lightning speed, but I could not care less about the watercolour painting of some glen or other. I see nothing but my autograph book on its display table with a sign in front of it. It bears the Lyon and Turnbull crest and the words *SOLD*.

'You should be celebrating, John!'

Mr Peebles is right, but I still walk in a daze, all the way back towards the station. It is just him and me now.

I imagine handing such a sum of money to Mother. I am a breadwinner without a job, but here is money we were never expecting! Instead of struggling to pay the rent, we will now be able to afford a deposit for a better house, perhaps the one in Priory Lane I saw advertised. It would be so handy for the town and for Mother's new factory job. It's at ground level and even has a small garden—Dey may be able to get outside again. A skip creeps into my step, all of its own accord. 'Thanks for coming with me, Mr Peebles, and for arranging it all. Truthfully, I would never have known where to start. I owe you a huge debt, sir.'

The librarian shuffles onto the Fife train ahead of me and makes his way to a compartment at the back. The carriages are quiet enough. He sinks into his seat with a groan. 'I am pleased to have been able to help, John. To tell you the truth, it was a welcome surprise for something to go right, after the week I have had.'

I look across sharply. *How selfish of me, to only think of myself and my own affairs!* 'I am very sorry to hear that, sir. What happened?'

The librarian rubs his back. 'Nothing much. Only I had to let my assistant go for carelessness, that's all. Enough was enough in the end. I have been on my own in the library all week though my sister is covering for me today. I am getting on a bit. I probably shouldn't be climbing those ladders at my age.'

We both stare into each other's eyes, realising, at the same time, the power of Provenance. *God helps those who help themselves.* A shiver of excitement travels up my back before settling with a prickle at the back of my neck.

The librarian breaks the silence first. 'Now that your work on the bridge has come to an end, I… I don't suppose you'd consider…' he begins with a croak, but he cannot finish, because I have thrown myself at his neck. I didn't mean to hug him. It just happened.

Our train rattles across the viaduct at Queensferry, and for a moment, my heart is in my mouth once more as the bridge's struts and girders fly past. Along the Queensferry cantilever, and then the central girder. The Inchgarvie cantilever. Another central girder. The northern cantilever and we rattle back into Fife and onto solid ground, the steam whistling in celebration. I open the window and look back: the still waters of the Forth mirror the blue sky above.

That was my bridge, carrying our steam train across the water with grace and ease! Mine! I am fit to burst with pride.

★★★

The very next Saturday I watch Mother sign the rent agreement for our new house in Priory Lane before setting off for my first day at the Carnegie Library.

Mother brushes my hair aside and buttons the top

button of my shirt. 'That's better, John. Can't have you disgracing the family, can we now.' Her eyes moisten when she sees the joy in my step as I thunder down the communal stair, accompanied by wee Helen's shrieks of 'Good luck, Johnny!' As soon as I am out in the street, I unbutton the top of my shirt at my throat again.

First, I walk, then I skip a couple of steps before bursting into a run—I simply cannot help it. Overhead, the gulls circle over my town and over the Forth where painters have already begun to paint the bridge red. They will continue to do this all year round.

Still running, I think of the dried nuts I scattered near Rusty's favourite tree and then of Cora's promise to call at the library later this afternoon to see how I am faring. I smile and then laugh as I run, ready to begin the first day of the rest of my life.

In the distance, the bridge playfully pokes at wisps of cloud.

I slow my pace as I enter the library.

'Shhh,' whispers Mr Peebles, waiting for me to catch my breath by the door. He pins a badge to my waistcoat.

John Nicol

Library Assistant

'Welcome,' he adds quietly and holds out his hand.

The End

Author's Note

I got married on Edinburgh's Royal Mile but celebrated beneath the Forth Bridge. The bridge was one of my first impressions of Scotland when I visited as a small child, and a regular date-destination for my husband and I while we studied in Edinburgh. Later, I would cross the bridge on the train to secure my very first teaching job, and then daily as I commuted to Fife for work.

Isn't it all the more impressive that the Forth Bridge was built in the Victorian era? To us, many of the methods used in its construction now seem outdated and dangerous, but we forget how radically advanced the process was at the time, how skilled the engineers, how progressive the technology. I am not technically minded, but as I was browsing through a photography book about Victorian Scotland, the faces of the briggers really captured my interest. Were there children involved? Given the Victorians' dodgy record on child labour, I suspected there would have been. Was there a book about the people who built the bridge?

I struck gold: a quick internet search pointed me to *The Briggers* by Elspeth Wills. I still remember reading the first few pages and realising I would need a highlighter—almost all of it felt relevant to what I was planning to write about.

I have long admitted it—I am no natural researcher. My go-to strategy is to find an expert and ask politely if they would allow me to speak to them about their pet subject. A group of Forth Bridge heritage enthusiasts called The Briggers handled the research for the book I had found, and they had also campaigned for a memorial for those killed during the construction. I contacted them and met retired engineer Frank Hay.

What an unstoppable force that man is! It's as if the term enthusiast was invented for him alone.

Me: Hi Frank, I read about John Nicol, the young boy who fell from the bridge and survived. Do we know any more about him? I'd like to base my book on him.

Frank: I'll look into it.

Shortly after:

Frank: Please find attached 10-page-document detailing everything you could possibly want to know, including copies of birth certificate, parents' marriage certificate, Scottish and Australian newspaper clippings, doctor's details, census data, genealogy etc

In constant dialogue with The Briggers group, I set to work. Although I didn't find an image of the real John Nicol, I felt I was beginning to get to know this young boy. It is possible he was a breadwinner due to his tricky family situation—the only logical way for him to be working on the bridge at such a young age, especially if the widow's fund had run out. Could he, like me, be encumbered by an irrational fear of heights? Endearingly, there are actual records of a squirrel on the bridge, even falling once and being fished out—Rusty was born in my imagination—who doesn't love an animal sidekick?

All the famous Victorians I mention really visited the bridge site—the only licence I took was to include Robert Louis Stevenson, one of my own favourite writers, who had in fact left Scotland for good the year before John starts work on the bridge. The Crown Prince, the golden rivet, the newspaper article about John's father's death, the riot at the Hawes Inn, Lady Moir who founded the Society of Women Engineers later—all of these are real.

The Carnegie Library had opened just a few years before the events described in *Rivet Boy*. The librarian was indeed a Mr Peebles, a bookbinder from Edinburgh—the opportunity to include a kindly librarian was too good to pass up. And in addition, the craze for autograph books had taken hold in America—a convenient way to weave in eminent Victorians. The story was beginning to come together.

All throughout the writing of the book, a shiny metal rivet kept me company on the shelf above my desk. Frank-the-Forth-Bridge-enthusiast gave it to me on my first research visit. It's poignant to me that rivets hold the bridge together, but that a rivet killed John's father. It seemed only right that a rivet has a part to play in the resolution of my story.

As for the title—I am normally terrible at choosing.
This time?
As Dey would say, *piece o cake!*

Acknowledgements

Once again, I must start by thanking Anne and Iain Glennie of Cranachan for believing in me and in this story, and for taking a chance on me once again. Without them, my writing would probably still languish in publishers' slush piles around the country.

Without the unwavering, unbridled enthusiasm of Frank Hay and the work of his fellow Briggers, this book simply wouldn't exist! I am also grateful for the support of the Forth Bridge Heritage Group.

In addition, I would like to thank the Dunfermline Carnegie Library & Galleries and the Hawes Inn for allowing me access to photograph the spaces I was writing about.

Thank you to my talented and faithful friend, the illustrator Sandra McGowan, for stunning chapter headings once again, and to Doda Smith for her picture editing skills.

Thanks to my fellow Clan Cranachan writers, to my fellow Time Tunnellers and my SCBWI tribe—thank you for always being in my corner.

Finally, and from the heart, thanks to Rob, Carla, Isla and Duncan—you're my gang, however hard the winds may blow. I thank God for you.

About the Author

Inverness-based Barbara Henderson is the author of historical novels *Fir for Luck*, *Punch*, *Black Water*, *The Siege of Caerlaverock* and *The Chessmen Thief* as well as the eco-thriller *Wilderness Wars*.

The Chessmen Thief won the Historical Association's Young Quills Award for the best children's historical fiction in 2022 and *The Siege of Caerlaverock* also won the same accolade in 2021.

Barbara shares her home with one teenage son, one long-suffering husband and a scruffy Schnauzer called Merry.

Web: barbarahenderson.co.uk
Twitter @scattyscribbler
Instagram: @scattyscribbler

A Picture is Worth a Thousand Words

'*A picture's worth a thousand words,*' Dey may say.
It is certainly true that contemporary photographs inspired much of *Rivet Boy's* story.

Many of the characters mentioned in the book really existed. Here are a few images which will allow you to visualise John Nicol's world: the Forth Bridge construction site, his beloved Carnegie Library in Dunfermline and some of the famous Victorians you have encountered in this story.

The Victorians loved this type of image: a Tableau Vivant or "living image", usually presented on a stage with static figures.

Here, some of the engineers demonstrate how a cantilever bridge works in a "human cantilever".

Kaichi Watanabe in the middle was one of the first Japanese engineers to train in Scotland and worked on the Forth Bridge.

BUILDING THE FORTH BRIDGE.

The towers of each cantilever were built up, and then out. Ships and boats patrolled the busy Forth beneath the structure.

South Cantilever, Forth Bridge. (23rd June 1888) 618.

Each cantilever was built separately like this.

This photo shows the Southern Cantilever, near where the site office and the Hawes Inn are located, just over two months before John begins to work there in *Rivet Boy*.

This etching shows the Crown Prince, Queen Victoria's son Prince Edward, then known as the Prince of Wales, securing the symbolic golden rivet to mark the completion of the Forth Bridge. William Arrol is assisting him, kneeling.

The Crown Prince became King Edward VII in January 1901, on the death of Queen Victoria.

The viaduct built to carry the train track towards the cantilevers of the bridge.

The Forth Bridge as seen from Hawes Pier in South Queensferry.

Briggers work high above the Forth, with minimal safety precautions. A thin wooden railing is all the protection they have. Tools were often dropped, putting those working at lower levels in danger.

Despite this, the workers seem at ease. There were no hard hats then!

South Queensferry Drill Works

The Hawes Inn is still trading today. Barbara took this picture on one of her research visits.

The two Forth Bridge designers, Benjamin Baker and John Fowler, with William Arrol.

William Arrol's company letterhead, Sir William Arrol Collection at Historic Environment Scotland, reproduced with kind permission.

A portrait of William Arrol, the Paisley-born contractor who not only built the Forth Bridge, but also the iconic Tower Bridge in London.

A remarkable man, he invented many of the tools used to make the construction process of his projects more efficient.

The famous Tower Bridge in London, also built by William Arrol.

At the time of construction between 22 April 1886 and 30 June 1894, Tower Bridge was the largest and most complex bascule bridge ever built ('bascule' comes from the French word for 'seesaw').

It was opened by the Prince and Princess of Wales with great festivities.

Margaret Moir, founder member of the *Women's Engineering Society*.

CC BY-SA 4.0 <https://creativecommons.org/licenses/by-sa/4.0>, via Wikimedia Commons

The philanthropist Andrew Carnegie, who used his immense wealth to do good—including funding 2811 free libraries around the world.

The entrance to the original Carnegie Library in Dunfermline. The building has now been extended to accommodate a wonderful museum and gallery, too.

The author on her very first visit to the bridge with her mother (standing), sister and brother-in-law in 1981. She is the child kneeling in front.